TESTING AND RELIABLE DESIGN OF CMOS CIRCUITS

TESTING AND RELIABLE DESIGN OF CMOS CIRCUITS

by

Niraj K. Jha
Princeton University

and

Sandip Kundu
IBM, T.J. Watson Research Center

KLUWER ACADEMIC PUBLISHERS
Boston/Dordrecht/London

Distributors for North America:
Kluwer Academic Publishers
101 Philip Drive
Assinippi Park
Norwell, Massachusetts 02061 USA

Distributors for all other countries:
Kluwer Academic Publishers Group
Distribution Centre
Post Office Box 322
3300 AH Dordrecht, THE NETHERLANDS

Library of Congress Cataloging-in-Publication Data

Jha, Niraj K.
    Testing and reliable design of CMOS circuits / by Niraj K. Jha
and Sandip Kundu.
      p.   cm. — (Kluwer international series in engineering and
computer science. VLSI, computer architecture, and digital signal
processing)
    ISBN 0-7923-9056-3
    1. Metal oxide semiconductors, Complimentary—Testing. 2. Metal
oxide semiconductors, Complimentary—Reliability. 3. Integrated
circuits—Very large scale integration—Design and construction.
I. Kundu, Sandip. II. Title. III. Series.
TK7871.99.M44J49   1990
621.39 '732—dc20   89-37031
                                                CIP

Printed in the United States of America.

To Our Parents

And

To Our Wives

TABLE OF CONTENTS

PREFACE

In the last few years CMOS technology has become increasingly dominant for realizing Very Large Scale Integrated (VLSI) circuits. The popularity of this technology is due to its high density and low power requirement. The ability to realize very complex circuits on a single chip has brought about a revolution in the world of electronics and computers. However, the rapid advancements in this area pose many new problems in the area of testing. Testing has become a very time-consuming process.

In order to ease the burden of testing, many schemes for designing the circuit for improved testability have been presented. These design for testability techniques have begun to catch the attention of chip manufacturers. The trend is towards placing increased emphasis on these techniques.

Another byproduct of the increase in the complexity of chips is their higher susceptibility to faults. In order to take care of this problem, we need to build fault-tolerant systems. The area of fault-tolerant computing has steadily gained in importance.

Today many universities offer courses in the areas of digital system testing and fault-tolerant computing. Due to the importance of CMOS technology, a significant portion of these courses may be devoted to CMOS testing. This book has been written as a reference text for such courses offered at the senior or graduate level. Familiarity with logic design and switching theory is assumed. The book should also prove to be useful to professionals working in the semiconductor industry.

Chapter 1 introduces the subject matter and deals with fault models, classification of CMOS circuits, etc. Chapter 2 points out some testing problems which are unique to CMOS circuits. Chapters 3 and 4 cover methods that can be used for testing dynamic and static CMOS circuits respectively. Chapter 5 presents various design for testability techniques. Chapter 6 deals with self-checking circuits. Many new error-detecting codes are also presented. Chapter 7 presents some conclusions. A comprehensive set of references is provided for those who wish to dig deeper into some particular area. Each chapter also includes many homework problems.

Acknowledgments: We would like to express our gratitude to our parents Dr. Chintamani Jha and Dr. Raj Kishori Jha, and Prof. Hari Mohan Kundu and Mrs. Prabhati Kundu for their constant encouragement and support. We would also like to thank our wives Shubha and Deblina without whose patience and understanding it would not have been possible to write this book.

Niraj K. Jha
Sandip Kundu

TESTING AND RELIABLE DESIGN OF CMOS CIRCUITS

Chapter 1
INTRODUCTION

Very Large Scale Integration (VLSI) has enabled us to implement very complex circuits on a single chip. Complementary Metal Oxide Semiconductor (CMOS) technology has played a dominant role in allowing this to happen. The advantages of VLSI circuits are obvious. However, they do pose a problem. The problem is how do we test the VLSI chips to ensure that they function as they are supposed to. With chips containing a million or more transistors, testing has become a difficult and time-consuming job. Testing is becoming an increasing part of the time it takes from conception to marketing of a chip. This problem can only become more severe in the future.

An approach which is advocated to ease the burden of testing is to use design for testability techniques. Many such techniques have been proposed. All of them have the common aim of trying to reduce the amount of time it takes to generate test vectors and apply them to the chip. Design for testability should become the rule rather than the exception in the future.

CMOS has been the dominant technology for the last few years and is expected to remain dominant for many years to come. However, CMOS poses many new challenges in the area of testing.

Overcoming these challenges is essential to the well-being of the semiconductor industry.

1.1 WHAT IS TESTING ?

Testing in the context of digital systems is defined to be the process by which a defect in the system can be exposed. The defect can occur at the time of manufacture or when the system is in the field. In Fig. 1.1 a device under test (DUT) is shown to which test vectors are applied. The resulting response from the device is monitored. If the correct response is known then we can determine if the DUT has a defect or not by comparing the responses.

Of course, it is assumed that one of the test vectors exposes the defect. For this assumption to be reasonable, the set of test vectors should include vectors for most, if not all, of the defects that are likely to occur.

1.2 FAULTS AND ERRORS

A fault is an actual defect that occurs in the device. When a vector is applied to the faulty device which produces an incorrect response, an error is said to have occurred. For example, if a line

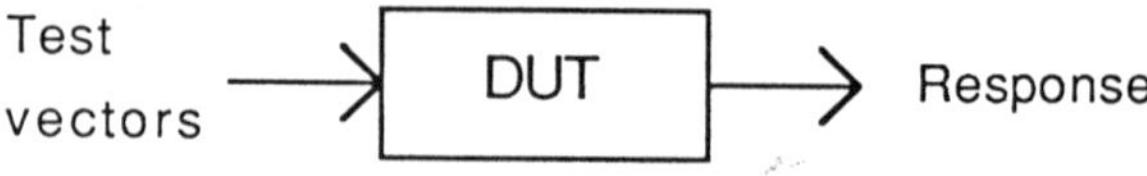

Fig. 1.1 Testing of a device

in a chip breaks, a fault has occurred. When this fault is exposed at the circuit outputs by some input vector, an error results. In this case the error is manifested as an incorrect logic value at one or more of the circuit outputs. However, monitoring the logic values is not the only way to determine if an error has occurred. There may be faults which cause the circuit to draw excessively large current when a particular input vector is applied, but they may not result in an incorrect logic value at the outputs. In this case the error is manifested as a drastic change in the value of the current.

A fault which can change the logic value on a line in the circuit from logic 0 to logic 1 or *vice versa* is called a logical fault. On the other hand if the fault causes some parameters of the circuit to change, such as the current drawn by the circuit, then it is termed parametric.

A fault can also be categorized on the basis of the duration for which it lasts. The three broad categories are (a) transient, (b) intermittent, and (c) permanent. A fault is called transient if it is only present for a small duration. A fault is intermittent if it appears regularly but is not present continuously. If a fault is present continuously, it is called permanent.

Transient faults have been the dominant cause of system failures [CAST82, IYER83]. These faults may be caused by α-particle radiation, power supply fluctuation, etc. No permanent damage is done by these faults. However, because of their short duration they are hard to detect. With the possibility of reduced voltage levels for VLSI and the resultant decrease in noise margins, system susceptibility to transient faults is likely to increase [YEN87].

Intermittent faults are also difficult to detect and locate [TASA77]. They can be caused by loose connections, bad designs or environmental effects like temperature and humidity variations. Permanent faults are the easiest to detect. They are predominantly caused by shorts and opens in VLSI circuits.

1.3 DIFFERENT TYPES OF CMOS CIRCUITS

Broadly speaking, one can categorize CMOS circuits as either static or dynamic. We discuss these categories next.

1.3.1 Static CMOS Circuits

A static CMOS circuit is made up of an interconnection of static CMOS gates, which consist of a network of pMOS transistors, called the pMOS network, and a network of nMOS transistors, called the nMOS network. The pMOS network is sometimes called the load network and the nMOS network is called the driver network [REDD84]. In positive logic a pMOS transistor conducts when its input is 0 and an nMOS transistor conducts when its input is 1.

Consider the CMOS gate in Fig. 1.2. Transistors 1 and 2 constitute the nMOS network, whereas transistors 3 and 4 constitute the pMOS network. Note the different symbols used for a pMOS transistor and an nMOS transistor. When the vector $(x_1,x_2) = 11$ is fed to the gate, both the nMOS transistors conduct and f is pulled down to ground (logic 0). For any other vector either one or both the transistors in the pMOS network conduct and f is pulled up to V_{dd} (logic 1). Therefore this gate acts like a NAND gate.

A static CMOS two-input NOR gate and an inverter are shown in Fig. 1.3. NAND, NOR and NOT (inverter) gates are referred to as primitive gates [El-ZI81]. Other CMOS gates, which

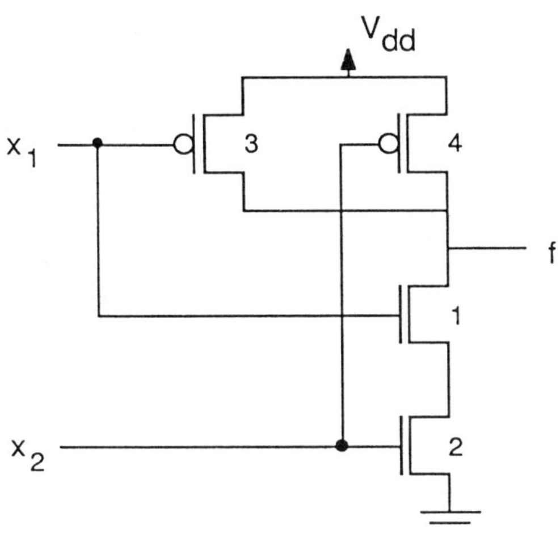

Fig. 1.2 A static CMOS gate

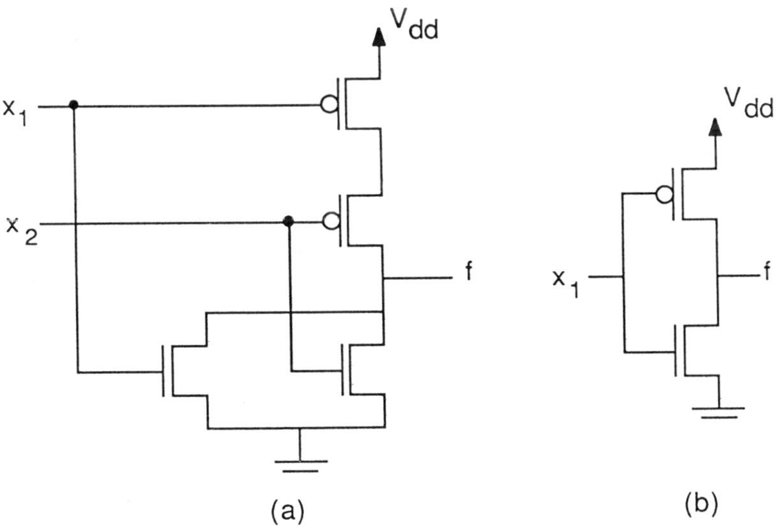

(a)   (b)

Fig. 1.3 (a) A static CMOS NOR gate (b) an inverter

have a more complex interconnection of transistors in their nMOS
and pMOS networks, are called complex gates. The CMOS gate

shown in Fig. 1.4 is an example of a complex gate. This complex gate implements the function $\mathrm{f} = \overline{\mathrm{x}_1\mathrm{x}_2 + \mathrm{x}_3\mathrm{x}_4}$.

The CMOS gates shown in Figs. 1.2, 1.3 and 1.4 all have the following two properties: (a) corresponding to each nMOS transistor there exists a pMOS transistor which is fed by the same input, and *vice versa*, (b) for each input vector a conduction path is activated either in the pMOS network or the nMOS network, but not both. Such CMOS gates are sometimes called fully complementary MOS (FCMOS) gates [REDD84]. If one of the networks is known to be series-parallel in nature then it is easy to derive the other network from it. This can be done by noting that a series (parallel) connection of transistors in the nMOS network corresponds to a parallel (series) connection of transistors in the

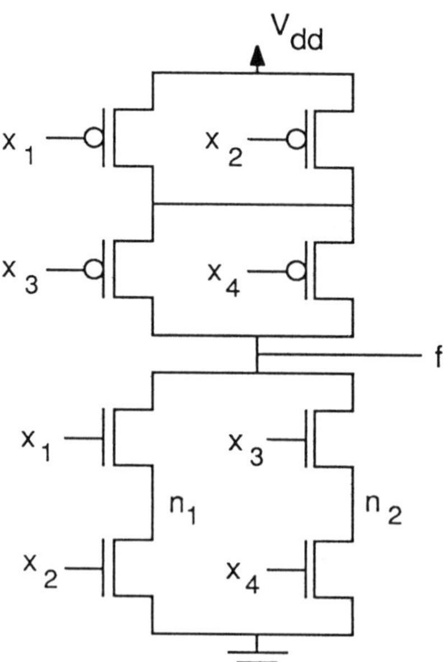

Fig. 1.4 A static CMOS complex gate

pMOS network. However, if one of the networks is known to be non-series-parallel in nature then it takes slightly more effort to derive the other network from it. This is best illustrated through an example.

Example 1.1: Consider the nMOS network shown in Fig. 1.5. In order to derive the pMOS network from it we first obtain the set of cutsets, which is explained next. A cutset is usually defined with respect to a graph. Imagine the graph in which an arc labeled by the transistor number replaces a transistor from the network in Fig. 1.5. A cutset of the graph is a set of branches whose removal will prevent any conduction between the two terminals (f and ground), whereas if any one branch from the cutset is added back to the graph then a conducting path is possible between the two terminals. Therefore the set of cutsets for the nMOS network can be derived as $\{(1,4), (1,3,5), (2,3,4,), (2,5)\}$. When there is no conduction in the nMOS network there is conduction in the pMOS network. Thus the set of cutsets given above define the set of conducting paths that are possible in the

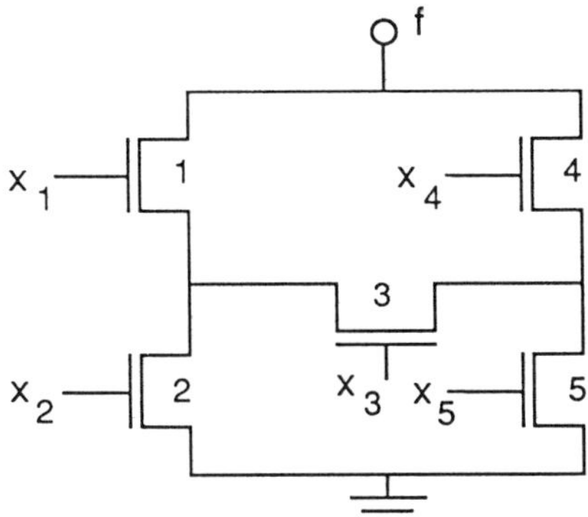

Fig. 1.5 A non-series-parallel nMOS network

pMOS network. This network is shown in Fig. 1.6. The set of conducting paths in this network is {(1,4), (1,3,5), (2,3,4), (2,5)} which is precisely the set of cutsets derived before. An nMOS and pMOS transistor which have the same number also have the same input. Similarly, an nMOS network can be derived, given a pMOS network. An efficient method for deriving cutsets has been given in [TSUK79]. □

Other types of static CMOS complex gates are also known which are not fully complementary. They violate property (a) but satisfy property (b) given before for FCMOS gates. Examples are Hybrid CMOS and PS-PS complex gates [JHA85, REDD86]. These will be discussed in later chapters.

1.3.2 Dynamic CMOS Circuits

Many schemes have been presented in literature for obtaining dynamic CMOS circuits [WEST85]. However, they all have some features in common. They have a precharge phase and an

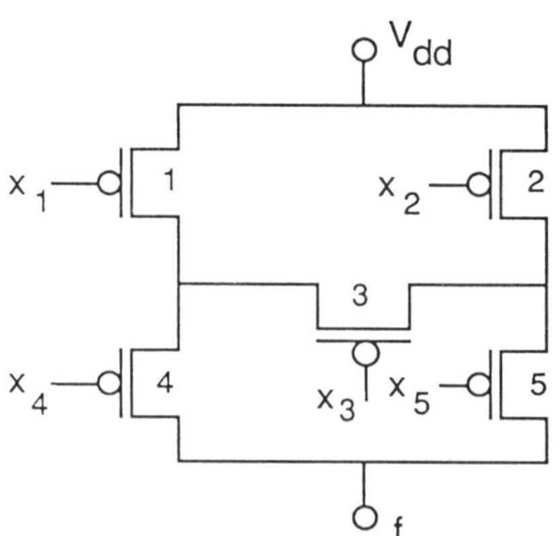

Fig. 1.6 The corresponding pMOS network

evaluation phase governed by one or more clocks. Consider the domino CMOS circuit [KRAM82] shown in Fig. 1.7. This circuit consists of a complex gate followed by an inverter and realizes the function $f = x_1x_2 + x_3x_4$. When clock $= 0$, the clocked pMOS transistor P1 conducts and precharges node g, while the conduction paths from g to ground are blocked by the off transistor N1. This is called the precharge phase and f attains the value 0 during this phase. When clock $= 1$ the circuit enters the evaluation phase. Transistor P1 is turned off and N1 is turned on. Depending on the input vector, the node g is either pulled down to 0 or remains at 1.

In a domino CMOS circuit every complex gate is always followed by an inverter. The outputs of these inverters in the circuit become 0 during the precharge phase. Thus the nMOS transistors

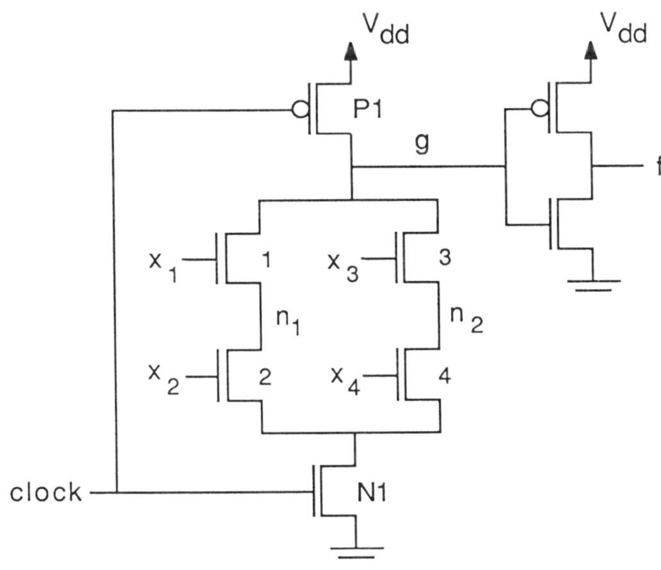

Fig. 1.7 A domino CMOS circuit

that they feed in other domino CMOS gates are turned off. Also, in a domino CMOS circuit the vectors at the primary inputs are changed only during the precharge phase. The above two facts ensure a glitch-free operation [KRAM82].

The advantage of domino CMOS over static CMOS is that it requires less area on the chip. Furthermore, since the pMOS network of a static CMOS gate is replaced by a single clocked pMOS transistor in a domino CMOS gate, the capacitive load at the output node becomes much less. Thus there is a potential for increasing the speed of the circuit. Another advantage is that a domino CMOS circuit is much more testable compared to a static CMOS circuit, as we will see in later chapters.

The disadvantage of a domino CMOS circuit is that it may require slightly more power than a static CMOS circuit. This is due to the fact that each domino CMOS gate must be precharged to logic 1 in every cycle even if its output is to continue to be logic 0. Another disadvantage of the domino CMOS technique is that it can only implement non-inverting functions. This is due to the fact that every domino CMOS gate has to be followed by an inverter and the output of any domino CMOS gate can not be fed directly to another gate. However, domino CMOS is fully compatible with static CMOS. Therefore it is possible to get around this problem by using static CMOS gates, when necessary, with the domino CMOS circuit.

Another way of getting rid of the above problem is by using differential cascode voltage switch (DCVS) logic circuits [HELL84]. The DCVS technique is an extension of the domino CMOS technique. However, a DCVS circuit can implement both inverting and non-inverting functions.

A two-input clocked DCVS EX-OR gate is shown in Fig. 1.8. The two inverters are, of course, static CMOS inverters. One can

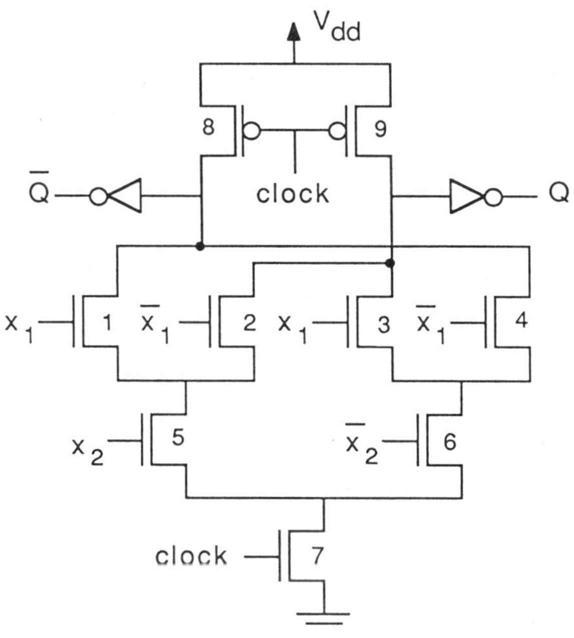

Fig. 1.8 A two-input DCVS EX-OR gate

see the similarities between this EX-OR gate and a domino CMOS gate. There are two pull-up clocked pMOS transistors instead of one. During precharge phase Q and $\overline{Q}$ become 0. In the evaluation phase the EX-OR function is realized at Q and the EX-NOR function at $\overline{Q}$. In some cases two feedback pMOS transistors are added to a DCVS gate to avoid the problem of charge sharing and leakage. This will be discussed in detail in Chapter 2.

If only Q or $\overline{Q}$ output, but not both, is desired then we can use a single-ended cascode voltage switch (SCVS) gate. For example, an SCVS EX-OR gate can be easily derived from the DCVS EX-OR gate in Fig. 1.8 by only including that part of the circuit which generates Q. This part consists of transistors 2, 3, 5, 6, 7, 9 and the inverter. An SCVS gate, thus, is essentially a domino

CMOS gate. A cascode voltage switch (CVS) logic circuit can be obtained by interconnecting DCVS and SCVS gates. However, we need to keep in mind that although a DCVS gate can feed an SCVS gate, an SCVS gate can not feed a DCVS gate. This is owing to the fact that a DCVS gate always needs both an input and its complement, which can not be supplied by an SCVS gate.

Just like a domino CMOS circuit, the operation of a CVS circuit is also glitch-free. This glitch-free operation gives dynamic CMOS circuits a significant testability advantage over static CMOS circuits, as we will see in the next chapter. A design technique for obtaining DCVS circuits can be found in [CHU86]. An algebraic technique given in [BRAY82] can also be used.

Many other variations of the domino CMOS and CVS techniques have also been presented in literature [FRIE84, PFEN85, PRET86, LEE86, GROT86, HWAN89]. A technique called NORA CMOS was presented in [GONC83] for implementing dynamic CMOS pipelined logic structures. Some other dynamic CMOS structures have been presented in [JI-RE87, LU88, KARL88, YUAN89]. In [CHU87] a nice comparison among the different dynamic CMOS circuit techniques can be found.

1.4 GATE-LEVEL MODEL

For testing a circuit, frequently its logic gate-level model is used. A gate-level model is simply the gate-level representation of the circuit.

Consider the static CMOS complex gate in Fig. 1.4. Its gate-level model is given in Fig. 1.9. This model is obtained by replacing a series connection of transistors in the nMOS network by an AND gate and a parallel connection by an OR gate. Since every CMOS gate is inverting, an inverter is placed at the output

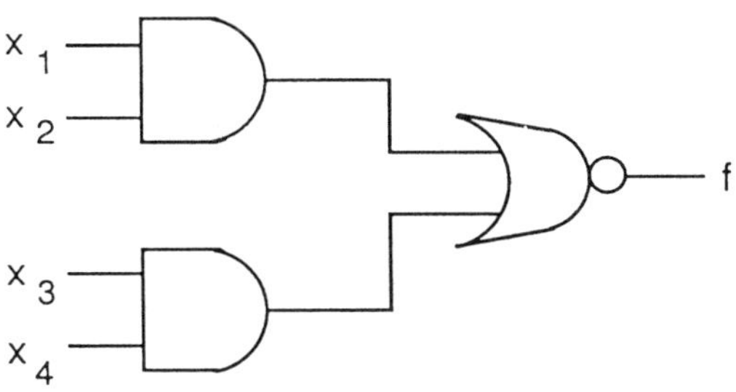

Fig. 1.9 The gate-level model of the circuit in Fig. 1.4

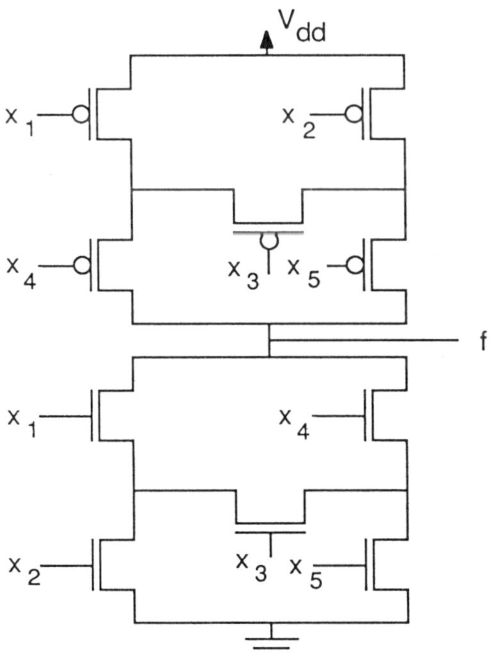

Fig. 1.10 A non-series-parallel static CMOS gate

of the model. If a CMOS circuit consists of many CMOS gates, each CMOS gate is replaced by its gate-level model and these are interconnected as in the original circuit to obtain the gate-level model of the whole circuit. A method, which derives the gate-level model from the pMOS network instead of the nMOS network of a CMOS gate, has been given in [REDD84].

The above scheme works well when the networks of the CMOS gates in the circuit consist of series and parallel connection of transistors. However, as mentioned earlier, it is also perfectly valid to connect the transistors in a non-series-parallel fashion, as shown in Fig. 1.10. This complex gate has been derived by combining the networks from Figs. 1.5 and 1.6. To derive a gate-level model for the circuit in Fig. 1.10 we first derive the set of loop-free conduction paths from the node f to ground [REDD84]. We can see that there are four paths in the nMOS network given by x_1x_2, $x_1x_3x_5$, x_4x_5 and $x_2x_3x_4$. Corresponding to each of these paths we have an AND gate, whose output is fed to a NOR gate. The gate-level model is shown in Fig. 1.11.

The gate-level model for a dynamic CMOS circuit can be similarly derived. For example, the gate-level model for the domino CMOS circuit in Fig. 1.7 is as shown in Fig. 1.12. This gate-level model is again derived from the nMOS logic network of the complex gate. For testing purposes, the clocked transistors need not be explicitly modeled at the gate-level. Faults in these transistors can easily be detected by test vectors derived from the gate-level model, as we will see in later chapters.

1.5 FAULT MODELS

If we were to try to derive test vectors for every possible physical failure in a VLSI chip, the problem would soon become unmanageable. In order to successfully tackle the problem, we represent the physical failures in a chip at a higher level with the help of a fault model [CASE76, TIMO83, HAYE85, SHEN85, ABRA86, FERG88].

Any one fault from the fault model may represent many physical failures. Thus the use of fault models speeds up the test generation process. The fault models most commonly used for CMOS circuits are (a) stuck-at, (b) stuck-open, (c) stuck-on, and (d) bridging. Another fault model which is increasingly being paid more attention is the delay fault model.

1.5.1 Stuck-at Fault Model

The fault model which has found the most widespread use in the industry is the stuck-at fault model. In this model it is assumed that the fault causes a line in the circuit to behave as if it is permanently at logic 0 or logic 1. If the line is permanently at logic 0 it is said to be stuck-at 0 (s-a-0), otherwise if it is permanently at logic 1 it is said to be stuck-at 1 (s-a-1).

Consider the two-input static CMOS NAND gate in Fig. 1.13. Let us first examine the short denoted by s_1. This short forces the

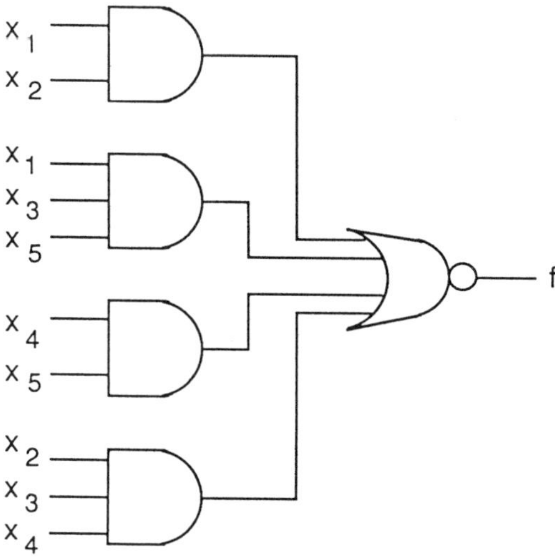

Fig. 1.11 The gate-level model of the circuit in Fig. 1.10

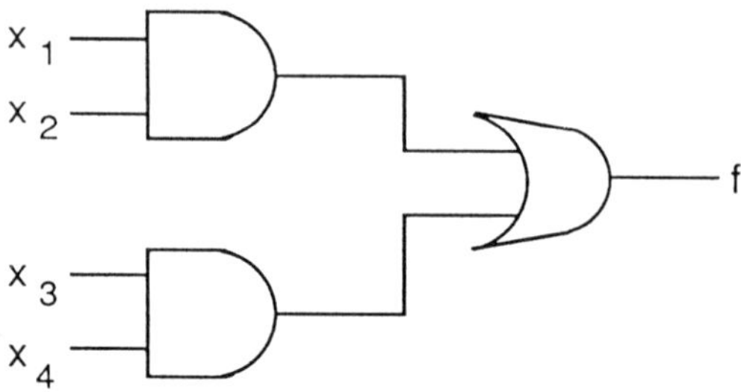

Fig. 1.12 The gate-level model for the circuit in Fig. 1.7

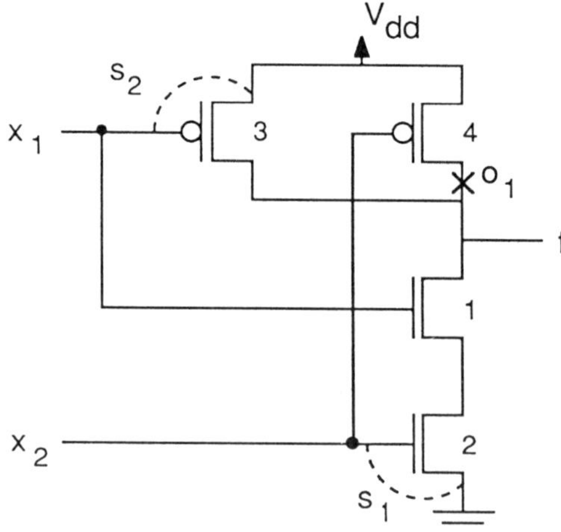

Fig. 1.13 A two-input CMOS NAND gate

line fed by input x_2 to behave in a s-a-0 fashion. Similarly, the short denoted by s_2 forces the line fed by input x_1 to behave in a s-a-1 fashion. In Table 1.1 the fault-free output is denoted as f whereas the outputs in the presence of shorts s_1 and s_2 are denoted as f_1 and f_2 respectively. From this table one can see that the vector $(x_1,x_2) = 11$ detects short s_1 and the vector $(x_1,x_2) = 01$ detects short s_2.

1.5.2 Stuck-open Fault Model

When a transistor is rendered non-conducting by a fault it is said to be stuck-open. The stuck-open fault model was first developed by Wadsack [WADS78]. Consider the break denoted by o_1 in Fig. 1.13. This break prevents transistor 4 from conducting. Thus it results in a stuck-open fault in transistor 4. Suppose that the vectors shown in Table 1.1 are applied in the order shown. Even when o_1 is present the resultant output will still be the same as the fault-free output f. This can be verified as follows. When 00 and 01 are applied, transistor 3 conducts, resulting in $f = 1$. When the third vector 10 is applied, neither the pMOS network nor the nMOS network can conduct. Therefore the previous logic

Table 1.1 Truth table for the NAND gate

x_1	x_2	f	f_1	f_2
0	0	1	1	1
0	1	1	1	0
1	0	1	1	1
1	1	0	1	0

value is retained at the output node. Finally, when 11 is applied, the nMOS network conducts and f becomes 0.

A stuck-open fault forces even a combinational CMOS circuit to behave in a sequential fashion. Thus in order to detect a stuck-open fault, a sequence of vectors is required. The reason the stuck-open fault in transistor 4 did not get detected above is that the proper sequence of vectors was not fed to the circuit.

It usually requires a sequence consisting of two vectors to detect a stuck-open fault. The first vector is called the initialization vector and the second vector is called the test vector. The sequence of these two vectors is referred to as the two-pattern test. The two-pattern test for the stuck-open fault in transistor 4 is $<11,10>$. The vector 11 initializes the output node to 0. When 10 is applied next, the output node remains at 0 and the fault is detected.

It should be mentioned that two-pattern tests should be applied at a rate more rapid than that associated with the leakage current time constants [WADS78]. Otherwise, a correct transition may be observed at the output even in the presence of the fault. This was also mentioned in [MALY88]. It was also pointed out in [MALY88], just as in [WADS78], that opens such as o_1 in Fig. 1.13 do not make the transistor permanently non-conducting. Due to the leakage currents the node f may eventually be charged to logic 1. However, the more important point is that if the two-pattern test is applied rapidly, the open o_1 will still be detected. So it is not necessary to assume that the transistor is permanently non-conducting for the success of two-pattern testing.

1.5.3 Stuck-on Fault Model

If a fault causes a transistor to conduct continuously, the transistor is said to be stuck-on [WADS78]. Consider the two-input CMOS NOR gate in Fig. 1.14. Suppose that transistor 2 is stuck-on. We could try to detect this fault by feeding the circuit with the vector $(x_1, x_2) = 00$. When this vector is fed, transistors 2, 3 and 4 conduct in the presence of the fault. Suppose the on-resistance of the pMOS transistors is R_p and that of the nMOS transistors is R_n. Then the effective voltage V_f at node f in the above situation is given by

$$V_f = \frac{R_n}{R_n + 2R_p} \times V_{dd}.$$

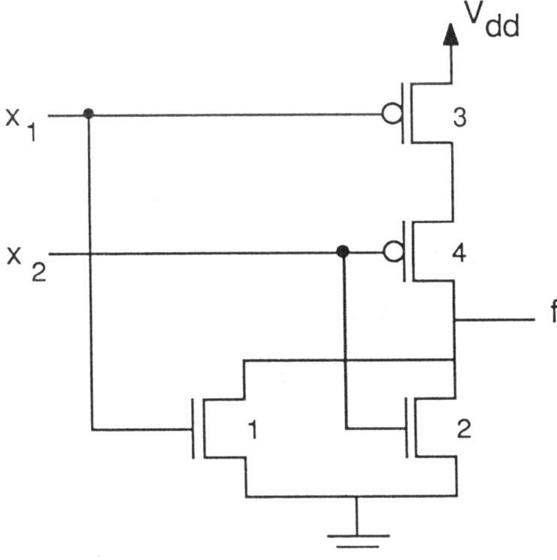

Fig. 1.14 A two-input static CMOS NOR gate

Since the off-resistance of transistor 1 would be much larger than the on-resistance of transistor 2, we have not taken the effect of transistor 1 into account. From the above equation we see that V_f can be anywhere between 0 volts and V_{dd} depending on what the relative values of R_n and R_p are. Thus there is no guarantee that the stuck-on fault will be detected if we only monitor the logic value at node f. We see that the only vector which has any chance of detecting the stuck-on fault is 00. Therefore changing the vector will not help.

Let us suppose for the sake of argument that V_f is such that it can be interpreted as logic 0. In that case the fault will be detected. Let us now switch our attention temporarily to the stuck-on fault in transistor 4. If this is the only fault present in the circuit then the only vector which has any chance of detecting it is 01. When 01 is fed, transistors 2, 3 and 4 conduct in the presence of the fault. Therefore the effective voltage at node f will be the same as V_f derived before. However, in order to detect this stuck-on fault in transistor 4 we need V_f to be interpreted as logic 1. Therefore the requirements for the detection of the stuck-on faults in transistors 2 and 4 are contradictory, and it is impossible to detect both the faults if we only monitor the logic value at f. Similarly, it is impossible to detect the stuck-on faults in both transistors 1 and 3. Thus detection of at most only two out of the four possible stuck-on faults is what we can expect.

In order to get around the above problem we can monitor the current drawn by the circuit instead of monitoring the logic at node f [LEVI81, MALA82, ACKE83, REDD84, MALA84]. When the circuit is fed the proper test vector, the amount of current drawn in the presence of a stuck-on fault is generally a few orders of magnitude greater than the normal leakage current [MALA82]. This is owing to the low-resistance conduction path that is activated between V_{dd} and ground. Thus current monitoring is a very effective technique for detecting stuck-on faults. The disadvantage of this technique is that it is slower than logic monitoring. However, we saw earlier that logic monitoring is totally inadequate even if we want to only detect all the single stuck-on faults in the circuit.

Another approach is advocated in [BASC86] for testing stuck-on faults. In this approach it is shown that stuck-on faults result in extra delay in the propagation of signals from primary inputs to circuit outputs. Thus by sampling the output at the proper time the fault is detected. However, this approach will work only if the logic value at the output node of the faulty CMOS gate is different from the fault-free case. But we saw earlier that this is not the case for many stuck-on faults. Therefore current monitoring is the only known technique which can be used for comprehensive testing of stuck-on faults.

1.5.4. Bridging Fault Model

A bridging fault is generally defined to be a short among two or more signal lines in the circuit. Such a short could occur, for example, due to defective masking or etching, aluminum migration, breakdown of insulators, etc [MEI74]. A bridging fault can be broadly classified as either (a) feedback bridging fault, or (b) non-feedback bridging fault. If a bridging fault creates one or more feedback loops, it is referred to as a feedback bridging fault, otherwise it is referred to as a non-feedback bridging fault [ABRA83].

Many researchers have looked into the problem of bridging fault detection in gate-level circuits [FRIE74, MEI74, IOSU78, KODA80, XU82, ABRA83, KARP83, YAMA84]. For gate-level circuits it is generally assumed that a bridging fault between two lines results in a wired-AND or wired-OR connection. This assumption has been found to be valid for ECL, RTL, DTL and TTL technologies [KODA80]. However, it is not valid for CMOS technology as we will see next.

Consider the circuit in Fig. 1.15. Suppose there is a bridging fault between lines f_1 and f_2 as shown by the dotted line. When the vector $(x_1,x_2,x_3) = 011$ is applied to the circuit, $f_1 = 1$ and $f_2 = 0$ in the fault-free case. However, when the bridging fault is present there is a low-resistance path from V_{dd} to ground through the pMOS network of the gate with output f_1 and the nMOS network of the gate with output f_2 [ACKE83]. Thus the voltage at

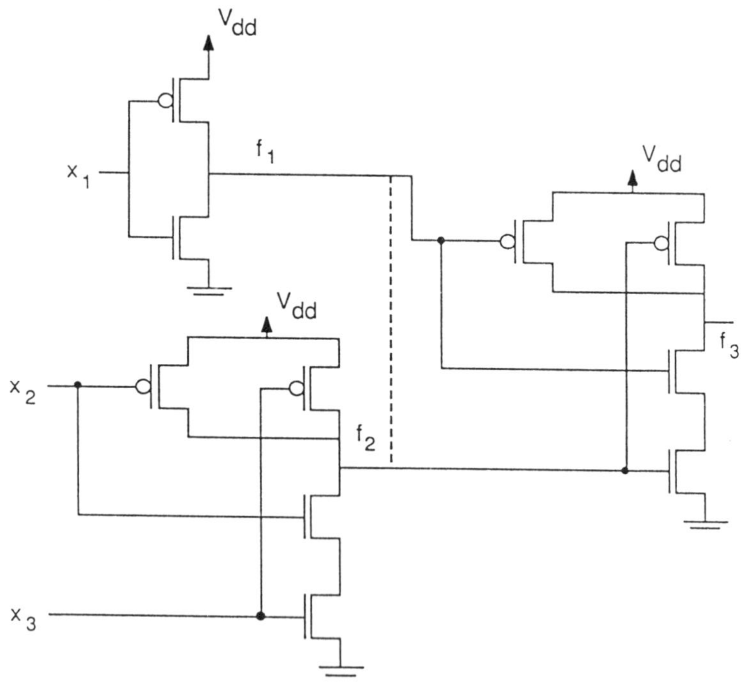

Fig. 1.15 A CMOS circuit

line f_1 or f_2 would be some intermediate value between V_{dd} and 0 volts. This situation is similar to the case of stuck-on faults except that the pMOS and nMOS networks under question belong to different CMOS gates. Depending on the value of the voltage at f_1 or f_2, it could be interpreted as logic 0 or 1 by the NAND gate they feed. There is another possibility. Suppose that line f_1 fans out to another CMOS gate with output f_4. Then it is possible for an intermediate voltage at f_1 to be interpreted as logic 0 by the gate with output f_3 and logic 1 by the gate with output f_4, or *vice versa*.

A more complex situation can also arise. Suppose for the sake of argument that when the vector 011 is fed to the CMOS circuit in Fig. 1.15, the voltage at f_1 and f_2 is such that it is interpreted as logic 1 by the NAND gate with output f_3. If we now apply the vector 101 to the circuit then a conduction path from V_{dd} to ground is activated through the pMOS network of the gate with output f_2 and the nMOS network of the gate with output f_1. In this case the voltage at f_1 and f_2 may well be such that it is interpreted as logic 0 by the subsequent NAND gate. Thus the same bridging fault could give rise to different logical behavior for different input vectors.

From the above discussions we can conclude that it is not easy to predict how a bridging fault will affect the logic values in a CMOS circuit. Clearly, a wired-AND or wired-OR model is not suitable for CMOS circuits. However, since there is a low-resistance path activated from V_{dd} to ground when a suitable vector is applied, current monitoring can easily detect the bridging fault [LEVI81, MALA82, ACKE83, MALA86].

1.5.5 Delay Fault Model

Even if a circuit is free of structural defects, it may not propagate a signal in the time allowed. This gives rise to a delay fault. The voltage on a faulty line could either be slow-to-rise (STR) or slow-to-fall (STF). We know from Section 1.5.3 that a stuck-on fault may give rise to a delay fault. Structural impurities may also result in such a fault.

Two types of delay fault model are generally used: (a) gate delay model, and (b) path delay model. The gate delay model models delay defects at the inputs or output of a gate. On the other hand, the path delay model models those defects which cause cumulative propagation delays along a circuit path to exceed the specified value. Each model has its own advantages and disadvantages. The path delay model requires the enumeration of circuit paths from all primary inputs to all circuit outputs. This causes an explosion in the number of paths that have to be considered, thereby increasing the number of tests and the test generation time. The gate delay model does not have this problem. However, it can not model delay defects which are not necessarily localized to single gates.

REFERENCES

[ABRA86] J. A. Abraham and W. K. Fuchs, "Fault and error models for VLSI," *Proc. IEEE*, vol. 74, no. 5, pp. 639-654, May 1986.

[ABRA83] M. Abramovici and P. R. Menon, "A practical approach to fault simulation and test generation for bridging faults," in *Proc. Int. Test Conf.*, Philadelphia, PA, pp. 138-142, Oct. 1983.

[ACKE83] J. M. Acken, "Testing for bridging faults (shorts) in CMOS circuits," in *Proc. Design Automation Conf.*, Miami Beach, FL, pp. 717-718, June 1983.

[BASC86] D. Baschiera and B. Courtois, "Advances in fault modeling and test pattern generation for CMOS," in *Proc. Int. Conf. Computer Design*, Port Chester, NY, pp. 82-85, Oct. 1986.

[BRAY82] R. K. Brayton and C. McMullen, "The decomposition and factorization of Boolean expressions," in *Proc. Int. Symp. Circuits & Systems*, Rome, Italy, pp. 49-54, June 1982.

[CASE76] G. R. Case, "Analysis of actual fault mechanisms in CMOS logic gates," in *Proc. Design Automation Conf.*, San Francisco, CA, pp. 265-270, June 1976.

[CAST82] X. Castillo, S. R. McConnel, and D. P. Siewiorek, "Derivation and calibration of a transient error reliability model," *IEEE Trans. Comput.*, vol. C-31, pp. 658-671, July 1982.

[CHU86] K. M. Chu and D. L. Pulfrey, "Design procedures for differential cascode voltage switch circuits," *IEEE J. Solid-State Circuits*, vol. SC-21, no. 6, pp. 1082-1087, Dec. 1986.

[CHU87] K. M. Chu and D. L. Pulfrey, "A comparison of CMOS circuit techniques: Differential cascode voltage switch logic versus conventional logic," *IEEE J. Solid-State Circuits*, vol. SC-22, no. 4, pp. 528-532, Aug. 1987.

[EL-ZI81] Y. M. El-Ziq and R. J. Cloutier, "Functional-level test generation for stuck-open faults in CMOS VLSI," in *Proc. Int. Test Conf.*, Philadelphia, PA, pp. 536-546, Oct. 1981.

[FERG88] F. J. Ferguson and J. P. Shen, "A CMOS fault extractor for inductive fault analysis," *IEEE Trans. CAD*, vol. 7, pp. 1181-1194, Nov. 1988.

[FRIE74]   A. D. Friedman, "Diagnosis of short-circuit faults in combinational circuits," *IEEE Trans. Comput.*, vol. C-23, pp. 746-752, July 1974.

[FRIE84]   V. Friedman and S. Liu, "Dynamic logic CMOS circuits," *IEEE J. Solid-State Circuits*, vol. SC-19, no. 2, pp. 263-266, Apr. 1984.

[GONC83]   N. F. Goncalves and H. J. De Man, "NORA: A racefree dynamic CMOS technique for pipelined logic structures," *IEEE J. Solid-State Circuits*, vol. SC-18, no. 3, pp. 261-266, June 1983.

[GROT86]   T. A. Grotjohn and B. Hoefflinger, "Sample-set differential logic (SSDL) for complex high-speed VLSI," *IEEE J. Solid-State Circuits*, vol. SC-21, no. 2, pp. 367-369, Apr. 1986.

[HAYE85]   J.P. Hayes, "Fault modeling," *IEEE Design & Test*, vol. 2, no. 2, pp. 88-95, Apr. 1985.

[HELL84]   L. G. Heller et al., "Cascode voltage switch logic: A differential CMOS logic family," in *Proc. Int. Solid-State Circuits Conf.*, pp. 16-17, Feb. 1984.

[HWAN89]   I. S. Hwang and A. L. Fisher, "Ultrafast compact 32-bit CMOS adders in multiple-output domino logic," *IEEE J. Solid-State Circuits*, vol. 24, no. 2, pp. 358-369, Apr. 1989.

[IOSU78]   A. Iosupovicz, "Optimal detection of bridging faults and stuck-at faults in two-level logic," *IEEE Trans. Comput.*, vol. C-27, no. 5, pp. 452-455, May 1978.

[IYER83]   R. K. Iyer and D. J. Rossetti, "Permanent CPU errors and systems activity: Measurement and modeling," in *Proc. Real-Time Syst. Symp.*, Arlington, VA, pp. 61-72, Dec. 1983.

[JHA85]   N. K. Jha and J. A. Abraham, "Design of testable CMOS circuits under arbitrary delays," *IEEE Trans. CAD*, vol. CAD-4, pp. 264-269, July 1985.

[JI-RE87]   Y. Ji-Ren, I. Karlsson, and C. Svensson, "A true single-phase-clock dynamic CMOS circuit technique," *IEEE J. Solid-State Circuits*, vol. SC-22, no. 5, pp. 899-901, Oct. 1987.

[KARL88] I. Karlsson, "True single phase clock dynamic CMOS circuit technique," in *Proc. Int. Symp. Circuits & Systems*, Espoo, Finland, pp. 475-478, June 1988.

[KARP83] M. Karpovsky, "Universal tests for detection of input/output stuck-at and bridging faults," *IEEE Trans. Comput.*, vol. C-32, no. 12, pp. 1194-1198, Dec. 1983.

[KODA80] K. L. Kodandapani and D. K. Pradhan, "Undetectability of bridging faults and validity of stuck-at fault test sets," *IEEE Trans. Comput.*, vol. C-29, no. 1, pp. 55-59, Jan. 1980.

[KRAM82] R. H. Krambeck, C. M. Lee, and H.-F. S. Law, "High-speed compact circuits with CMOS," *IEEE J. Solid-State Circuits*, vol. SC-17, no. 3, pp. 614-619, June 1982.

[LEE86] C. M. Lee and E. W. Szeto, "Zipper CMOS," *IEEE Circuits & Devices*, pp. 10-17, May 1986.

[LEVI81] M. W. Levi, "CMOS is most testable," in *Proc. Int. Test Conf.*, Philadelphia, PA, pp. 217-220, Oct. 1981.

[LU88] S.-H. Lu, "Implementation of iterative networks with CMOS differential logic," *IEEE J. Solid-State Circuits*, vol. 23, no. 4, pp. 1013-1017, Aug. 1988.

[MALA82] Y. K. Malaiya and S. Y. H. Su, "A new fault model and testing technique for CMOS devices," in *Proc. Int. Test Conf.*, Philadelphia, PA, pp. 25-34, Oct. 1982.

[MALA84] Y. K. Malaiya, "Testing stuck-on faults in CMOS integrated circuits," in *Proc. Int. Conf. Computer-Aided Design*, Santa Clara, CA, pp. 248-250, Nov. 1984.

[MALA86] Y. K. Malaiya, A. P. Jayasumana, and R. Rajsuman, "A detailed examination of bridging faults," in *Proc. Int. Conf. Computer Design*, Port Chester, NY, pp. 78-81, Oct. 1986.

[MALY88] W. Maly, P. K. Nag, and P. Nigh, "Testing oriented analysis of CMOS ICs with opens," in *Proc. Int. Conf.*

Computer-Aided Design, Santa Clara, CA, pp. 344-347, Nov. 1988.

[MEI74] K. C. Y. Mei, "Bridging and stuck-at faults," *IEEE Trans. Comput.*, vol. C-23, no. 7, pp. 720-727, July 1974.

[PFEN85] L. C. Pfennings et al., "Differential split-level CMOS logic for subnanosecond speeds," *IEEE J. Solid-State Circuits*, vol. SC-20, pp. 1050-1055, Oct. 1985.

[PRET86] J. A. Pretorius, A. S. Shubat, and C. A. T. Salama, "Latched domino CMOS logic," *IEEE J. Solid-State Circuits*, vol. SC-21, no. 4, pp. 514-522, Aug. 1986.

√[REDD84] S. M. Reddy, V. D. Agrawal, and S. K. Jain, "A gate-level model for CMOS combinational logic circuits with application to fault detection," in *Proc. Design Automation Conf.*, Albuquerque, NM, pp. 504-509, June 1984.

[REDD86] S. M. Reddy and M. K. Reddy, "Testable realizations for FET stuck-open faults in CMOS combinational logic circuits," *IEEE Trans. Comput.*, vol. C-35, pp. 742-754, Aug. 1986.

[SHEN85] J. P. Shen, W. Maly, and F. J. Ferguson, "Inductive fault analysis of MOS integrated circuits," *IEEE Design & Test*, vol. 2, no. 6, pp. 13-26, Dec. 1985.

[TASA77] O. Tasar and V. Tasar, "A study of intermittent faults in digital computers," in *Proc. AFIPS Conf.*, pp. 807-811, 1977.

[TIMO83] C. Timoc et al., "Logical models of physical failures," in Proc. *Int. Test Conf.*, Philadelphia, PA, pp. 546-553, Oct. 1983.

[TSUK79] S. Tsukiyama, H. Ariyoshi, and I. Shirakawa, "Algorithm to enumerate all the cutsets in $O(|V|+|E|)$ time per cutset," in *Proc. Int. Symp. Circuits & Systems*, Tokyo, Japan, pp. 645-648, June 1979.

[WADS78] R. L. Wadsack, "Fault modeling and logic simulation of CMOS and MOS integrated circuits," *Bell Syst. Tech. J.*, vol. 57, no. 5, pp. 1449-1474, May-June 1978.

[WEST85] N. Weste and K. Eshraghian, Principles of CMOS VLSI design: A systems perspective, Addison-Wesley, Reading, Mass., 1985.

[XU82] S. Xu and S. Y. H. Su, "Testing feedback bridging faults among internal, input and output lines by two patterns," in *Proc. Int. Conf. Circuits & Computers*, New York, NY, pp. 214-217, Oct. 1982.

[YAMA84] T. Yamada and T. Nanya, "Stuck-at fault tests in the presence of undetectable bridging faults," *IEEE Trans. Comput.*, vol. C-33, no. 8, pp. 758-761, Aug. 1984.

[YEN87] M. Y. Yen, W. K. Fuchs, and J. A. Abraham, "Designing for concurrent error detection in VLSI: Application to a microprogram control unit," *IEEE J. Solid-State Circuits*, vol. SC-22, no. 4, pp. 595-605, Aug. 1987.

[YUAN89] J. Yuan and C. Svensson, "High-speed CMOS circuit technique," *IEEE J. Solid-State Circuits*, vol. 24, no. 1, pp. 62-70, Feb. 1989.

PROBLEMS

1.1. For the two-input static CMOS NAND gate shown in Fig. 1.13, derive a test vector or a two-pattern test for the following faults:

    (a) a stuck-at 1 fault on the line fed by input x_2,

    (b) a stuck-open fault in transistor 1,

    (c) a stuck-on fault in transistor 4 assuming that current monitoring is done.

1.2. Derive the set of all the test vectors which can detect the bridging fault shown in Fig. 1.15, assuming that the current drawn by the circuit is monitored.

1.3. Obtain a test vector which would detect a short between nodes n_1 and n_2 in Fig. 1.4. If we just monitor the logic value at f, can this short be guaranteed to be detected by the vector that was derived ? Explain.

1.4. Obtain a gate-level model for the circuit given in Fig. 1.15. You can assume that the bridging fault shown in the circuit is not present. From the gate-level model find a test vector for detecting a stuck-at 0 fault on the line fed by x_2.

1.5. In the gate-level model derived in Problem 1.4, indicate which stuck-at faults will be detected by the test vector 001.

1.6. For the domino CMOS circuit in Fig. 1.7 obtain test vectors for the following faults:

(a) transistor 1 stuck-on,

(b) line x_4 stuck-at 1.

1.7. **(a)** Do we need a two-pattern test to detect the stuck-open fault in transistor 3 in the circuit in Fig. 1.7? Explain.

(b) What are the stuck-open faults in this circuit which require two-pattern tests ?

1.8. Obtain a test vector to detect the following faults in the DCVS EX-OR gate shown in Fig. 1.8:

(a) a stuck-open fault in the transistor fed by input x_2,

(b) a stuck-on fault in the transistor fed by input $\overline{x}_2$.

Chapter 2
TEST INVALIDATION

CMOS has become a very popular technology because of its low-power requirement and high density. However, an increase in the density of the chips also increases the complexity of testing. To further add to the woes of a testing engineer, new mechanisms have been identified through which a test derived for a CMOS circuit may be invalidated. In other words, a test may not be able to do its intended job. This is the topic of discussion in this chapter.

2.1 THE TEST INVALIDATION PROBLEM

Two mechanisms have been identified which can prevent a test for a CMOS circuit from detecting all the faults it is intended to detect. These mechanisms are circuit delays and charge sharing. We discuss them next.

2.1.1 Test Invalidation due to Circuit Delays

When the vector at the primary inputs of the circuit changes to another vector, the logic values at the inputs of a gate embedded in the circuit do not change simultaneously since different parts of the circuit have different delays. These circuit delays may prevent a two-pattern test from detecting the stuck-open fault it is devised to detect unless it is carefully derived [JAIN83, REDD83]. Similarly, timing skews among the primary inputs can also lead to test invalidation. A test which can not be invalidated is called robust, else it is called non-robust. It is known that for some circuits a robust test set does not exist [REDD83].

Consider the static CMOS circuit in Fig. 2.1, which realizes the function $f = \bar{x}_1 x_2 + x_1 \bar{x}_2 + \bar{x}_3 \bar{x}_4 + x_3 x_4 + \bar{x}_1 x_3$. Since this complex gate is derived from a sum-of-products expression, it is called an AND-OR CMOS realization. Suppose the pMOS transistor fed by x_1, as shown in Fig. 2.1, is stuck-open (s-op). We need a two-pattern test to detect this fault. The initialization vector should make $f = 0$. For this function only three vectors result in $f = 0$. These are $(x_1, x_2, x_3, x_4) = \{0001,1101,1110\}$. The test vector should try to activate a conduction path from V_{dd} to f through the faulty transistor. Only one such vector exists, which is 0010. Therefore there are three two-pattern tests possible: $<0001,$ $0010>$, $<1101, 0010>$, $<1110, 0010>$.

If we choose $<0001, 0010>$, then either 0000 or 0011 could be produced as an intermediate vector if the inputs x_3 and x_4 change at slightly different times due to timing skews at the primary inputs. Either of the two intermediate vectors would activate a conduction path in the pMOS network which is in parallel to the one containing the faulty transistor. Thus for the test vector, f will become 1 and the stuck-open fault will not be detected. Similarly, the other two two-pattern tests can also be invalidated by timing skews. Therefore for this stuck-open fault a robust two-pattern test does not exist. This in turn implies that the AND-OR CMOS realization in Fig. 2.1 is not robustly testable. Thus, another realization, which is robustly testable, has to be found for this function. This will be discussed in Chapter 5.

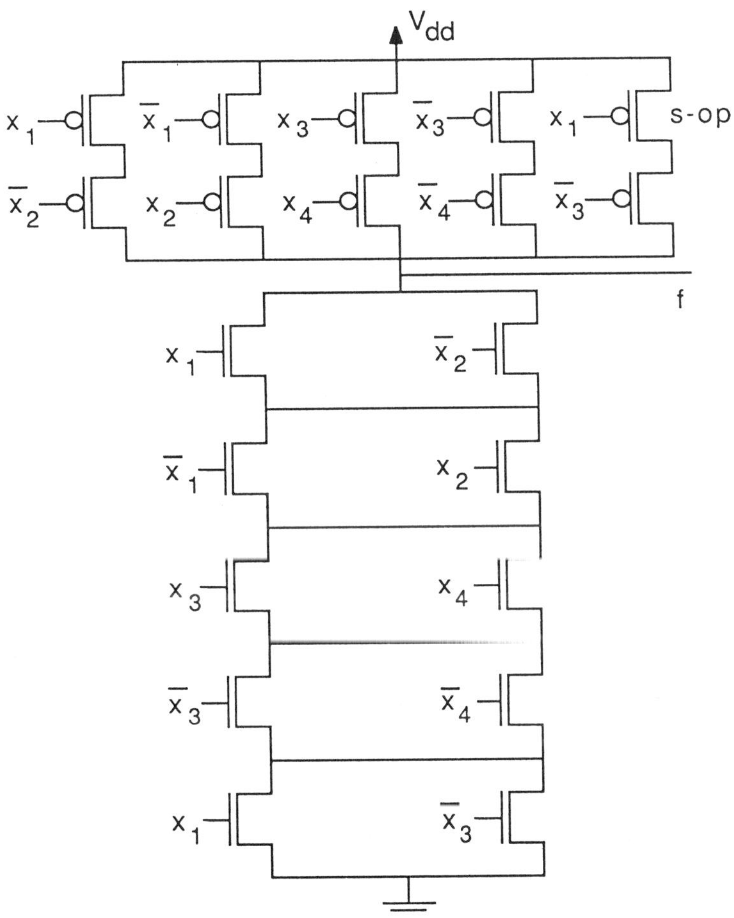

Fig. 2.1 A static CMOS AND-OR realization

The concept of robust testing is also applicable to delay
faults. A delay fault is said to be robustly testable if it can be

detected regardless of other delays in the circuit.

2.1.2 Test Invalidation due to Charge Sharing

Another mechanism through which a test can be invalidated is charge sharing between the output node of a faulty CMOS gate and its internal nodes [REDD86].

This mechanism is best illustrated through an example. We use an example which has been adapted from [REDD86]. Consider the static CMOS complex gate in Fig. 2.2. Suppose we want to test for a stuck-open fault in the nMOS transistor numbered 3. A possible two-pattern test for this fault is $<0101, 0001>$. The initialization vector 0101 initializes nodes m_5 through m_{11} to logic 1. The logic values at nodes m_1, m_2, m_3, m_4 and m_{12} remain undetermined since pMOS transistors 1, 3, 4, 6 and nMOS transistors 7 through 12 are non-conducting. A previous vector could determine the states of these nodes. Suppose that the vector 0011 was applied before the initialization vector. This vector would set nodes m_2, m_3, m_4 and m_{12} to logic 0. These nodes could retain the logic 0 value even when the initialization vector is applied. After that when the test vector 0001 is applied, the output node m_9 has no conduction path to ground due to the stuck-open fault in nMOS transistor 3. However, the pMOS transistors 2, 3, 5 and 6 conduct in the presence of the test vector. Therefore node m_9 will have to share its charge with nodes m_1 through m_4. Thus it is possible that the voltage at m_9 could become less than what can be recognized as logic 1, thereby invalidating the two-pattern test.

The effect of charge sharing on a test can be seen to be dependent on the manner in which the transistors in the pMOS or nMOS network are connected to each other. For example, if the inputs to pMOS transistors 1 and 4 are changed to x_2 and the

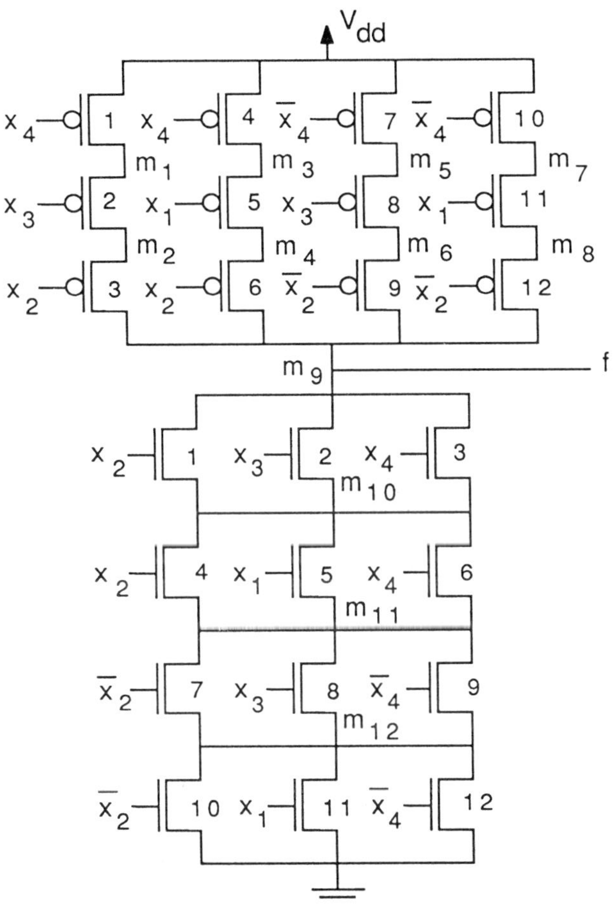

Fig. 2.2 A static CMOS complex gate

inputs to pMOS transistors 3 and 6 are changed to x_4, then the charge sharing illustrated earlier does not take place.

The example given here took into account charge sharing that occurs in the steady state. However, the problem could also arise in the transient state when one vector is changing to another. But the effect of the intermediate vectors would be less pronounced.

In practice, test invalidation due to charge sharing may be rare due to the following reasons:

(1)   Technological constraints usually limit the fan-in of a CMOS gate to 4-8.

(2)   The capacitance associated with the output node of a CMOS gate is usually much greater than the capacitance associated with internal nodes.

Owing to the above reasons the problem of test invalidation due to charge sharing has received much less attention than the problem of test invalidation due to circuit delays in static CMOS circuits.

2.2 ROBUST TESTABILITY OF DYNAMIC CMOS CIRCUITS

Dynamic CMOS circuits are inherently protected against test invalidation due to circuit delays. With the addition of a few extra transistors, a dynamic CMOS circuit can also be protected against test invalidation due to charge sharing.

Consider the domino CMOS [KRAM82] circuit in Fig. 2.3. Suppose we want to test for a stuck-open fault in nMOS transistor 1. A possible test vector is $(x_1, x_2, x_3, x_4, x_5, x_6, x_7) = 1110001$. This vector arrives at the primary inputs during the precharge phase. The evaluation phase starts only after the inputs have settled down. Note that a two-pattern test is not needed for detecting this fault since the initialization is automatically taken care of in the precharge phase. Suppose that the previous vector that was applied to the circuit was 0111110. Thus a possible intermediate vector is 1111111. However, since the intermediate vector can only occur during the precharge phase when all the clocked nMOS

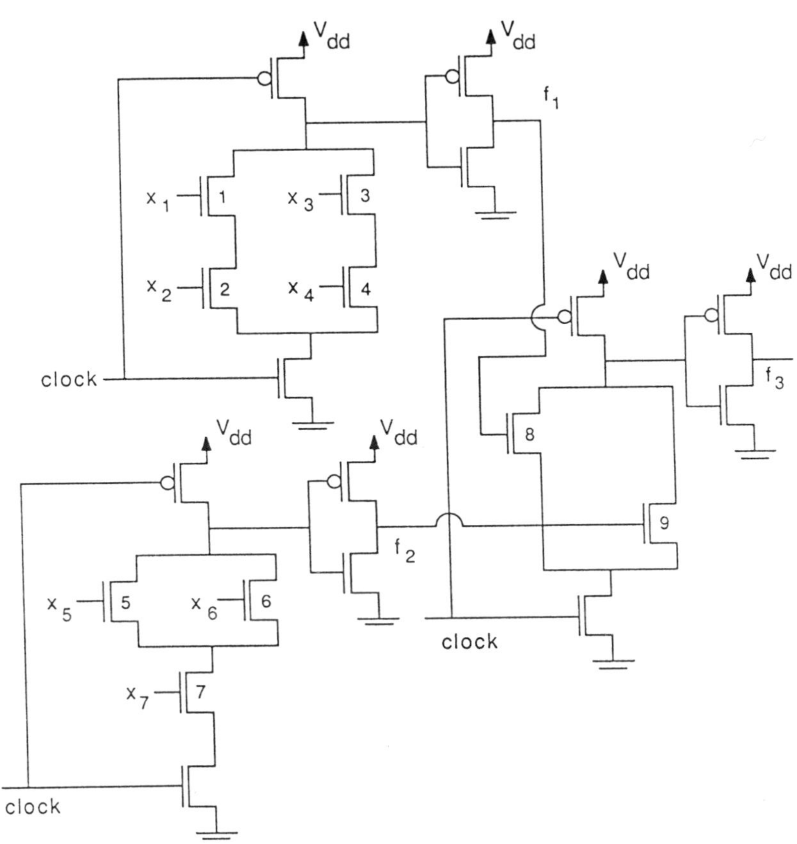

Fig. 2.3 A domino CMOS circuit

transistors are non-conducting, the test can not be invalidated.

Let us now consider a transistor which is not fed by a primary input. Suppose we want to test for a stuck-open fault in nMOS transistor 9. A possible test vector for this fault is 0101101. This vector results in $(0,1)$ at (f_1,f_2) during the evaluation phase. In the precharge phase just preceding this evaluation phase, (f_1,f_2) has $(0,0)$. Thus, no matter what the previous vector at the primary inputs was, only a $0 \rightarrow 1$ transition is possible at the inputs of transistors 8 and 9. Therefore a spurious conduction path can not be activated through transistor 8 and the stuck-open fault in transistor 9 is robustly detected.

We thus see that any test derived for a stuck-open fault in a domino CMOS circuit is inherently robust. The operation of a cascode voltage switch (CVS) logic circuit is very similar to the operation of a domino CMOS circuit. We can use similar arguments to show that a test for a CVS circuit is also inherently robust.

If some extra transistors are provided in dynamic CMOS circuits then the charge sharing problem can also be overcome. Consider the subcircuit whose output is f_1 in the circuit in Fig. 2.3. This subcircuit can be modified as shown in Fig. 2.4. A weak feedback pMOS transistor has been added in this circuit whose input

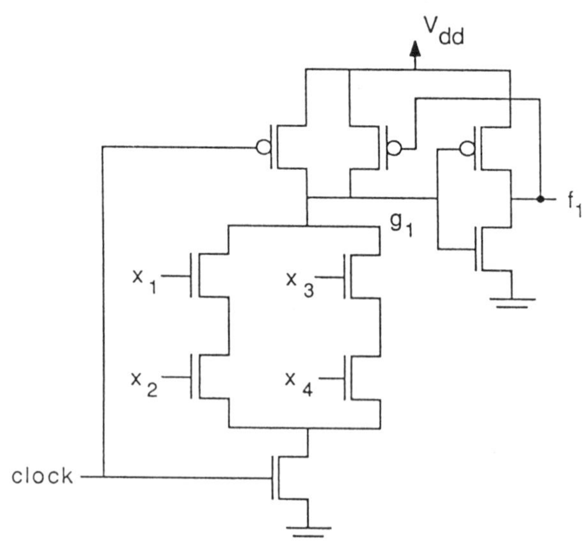

Fig. 2.4 The modified subcircuit

is controlled by the output of the inverter [HELL84, WEST85]. Due to charge sharing or leakage if node g_1 begins to lose its charge then the feedback transistor pulls it back to logic 1. Since this feedback transistor is weak, it does not fight the pull-down transistors in the nMOS network. In a domino CMOS circuit consisting of many gates, one feedback pMOS transistor can be added to every complex gate. The disadvantage of adding these transistors is that the speed of the circuit may be adversely affected.

A differential cascode voltage switch (DCVS) logic circuit can be similarly modified by adding two extra transistors to every complex gate in the circuit. For example, a modified two-input DCVS EX-OR gate is shown in Fig. 2.5.

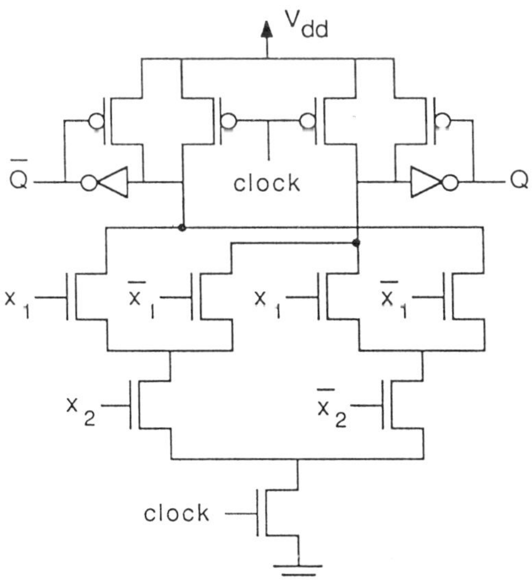

Fig. 2.5 A modified two-input DCVS EX-OR gate

REFERENCES

[HELL84] L. G. Heller et al., "Cascode voltage switch logic: A differential CMOS logic family," in *Proc. Int. Solid-State Circuits Conf.*, pp. 16-17, Feb. 1984.

[JAIN83] S. K. Jain and V. D. Agrawal, "Test generation for MOS circuits using D-algorithm," in *Proc. Design Automation Conf.*, Miami Beach, FL, pp. 64-70, June 1983.

[KRAM82] R. H. Krambeck, C. M. Lee, and H.-F. S. Law, "High-speed compact circuits with CMOS," *IEEE J. Solid-State Circuits*, vol. SC-17, no. 3, pp. 614-619, June 1982.

[REDD83] S. M. Reddy, M. K. Reddy, and J. G. Kuhl, "On testable design for CMOS logic circuits," in *Proc. Int. Test Conf.*, Philadelphia, PA, pp. 435-445, Oct. 1983.

[REDD86] S. M. Reddy and M. K. Reddy, "Testable realizations for FET stuck-open faults in CMOS combinational logic circuits," *IEEE Trans. Comput.*, vol. C-35, pp. 742-754, Aug. 1986.

[WEST85] N. Weste and K. Eshraghian, Principles of CMOS VLSI design: A systems perspective, Addison-Wesley, Reading, Mass., 1985.

ADDITIONAL READING

[AGRA84] P. Agrawal, "Test generation at the switch level," in *Proc. Int. Conf. Computer-Aided Design*, Santa Clara, CA, pp. 128-130, Nov. 1984.

[GUPT88] G. Gupta and N. K. Jha, "A universal test set for CMOS circuits," *IEEE Trans. CAD*, vol. 7, pp. 590-597, May 1988.

[JHA85] N. K. Jha and J. A. Abraham, "Design of testable CMOS circuits under arbitrary delays," *IEEE Trans. CAD*, vol.

CAD-4, pp. 264-269, July 1985.

[KUND88a] S. Kundu and S. M. Reddy, "On the design of robust testable combinational logic circuits," in *Proc. Int. Symp. Fault-Tolerant Comput.*, Tokyo, Japan, pp. 220-225, June 1988.

[KUND88b] S. Kundu, S. M. Reddy, and N. K. Jha, "On the design of robust multiple fault testable CMOS combinational logic circuits," in *Proc. Int. Conf. Computer-Aided Design*, Santa Clara, CA, pp. 240-243, Nov. 1988.

[MORI86] P. S. Moritz and L. M. Thorsen, "CMOS circuit testability," *IEEE J. Solid-State Circuits*, vol. SC-21, no. 2, pp. 306-309, Apr. 1986.

[REDD84] S. M. Reddy, M. K. Reddy, and V. D. Agrawal, "Robust tests for stuck-open faults in CMOS combinational circuits," in *Proc. Int. Symp. Fault-Tolerant Comput.*, Orlando, FL, pp. 44-49, June 1984.

[SHER88] S. D. Sherlekar and P. S. Subramanian, "Conditionally robust two-pattern tests and CMOS design for testability," *IEEE Trans. CAD*, vol. 7, pp. 325-332, Mar. 1988.

PROBLEMS

2.1. Suppose that the function realized by the circuit in Fig. 2.1 is instead implemented using a two-level NAND-NAND CMOS circuit. Indicate the stuck-open fault(s) in this circuit which can not be robustly detected. Explain.

2.2. Is the OR-AND CMOS realization based on the product-of-sums expression of the function realized in Fig. 2.1 robustly testable? If so, derive the robust test set, else point out the fault(s) which is not robustly testable.

2.3. Find a function of four variables, other than the one realized in Fig. 2.1, whose AND-OR CMOS realization is not robustly testable.

2.4. Find a function of five variables whose AND-OR and OR-AND CMOS realizations are both not robustly testable.

2.5. In the static CMOS complex gate in Fig. 2.2, if the pMOS transistors 1 and 4 are fed x_2 and the pMOS transistors 3 and 6 are fed x_4, show that the two-pattern test $<0101, 0001>$ discussed in Section 2.1.2 can not be invalidated by charge sharing. For this modified circuit can you find some other two-pattern test which may be invalidated.

Chapter 3
TEST GENERATION FOR DYNAMIC CMOS CIRCUITS

The dynamic CMOS circuit techniques present us with an attractive alternative to static CMOS circuits. It is possible to achieve higher density and speed with these techniques at the expense of a slight increase in the power requirement. In this chapter we will explore the testability of combinational dynamic CMOS circuits.

A test set can be derived for a CMOS circuit either directly from its switch-level description or from its gate-level model. We will use both these approaches.

The advantage of using a gate-level model is that many of the existing gate-level test generation techniques can be employed. Some of these techniques are path sensitization and D-algorithm [ARMS66, ROTH66], Boolean difference [SELL68], PODEM [GOEL81], FAN [FUJI83], SOCRATES [SCHU88]. A detailed description of all these techniques is beyond the scope of this book. However, we will describe here the path sensitization technique and its generalization called D-algorithm, which have found widespread use. We will also discuss the Boolean difference technique.

3.1 PATH SENSITIZATION AND D-ALGORITHM

Test generation techniques based on the gate-level model are generally geared towards detecting single stuck-at faults, although they can be extended to other fault models as well. If a circuit has n lines, it can have only 2n single stuck-at faults, whereas the number of possible multiple stuck-at faults is $3^n - 1$. Thus it is not possible to generate tests for each multiple stuck-at fault even for moderate values of n. A test set derived for single stuck-at faults also detects many multiple stuck-at faults. Therefore by applying such a test set to the circuit, the testing engineer hopes to catch a multiple stuck-at fault if such a fault is present in the circuit.

The path sensitization technique involves sensitizing a path through the circuit by assigning proper logic values to the gates on the path such that the effect of the fault is observed at the circuit output [ARMS66]. We can illustrate this technique with an example. Consider the circuit in Fig. 3.1. Suppose there is a stuck-at 0

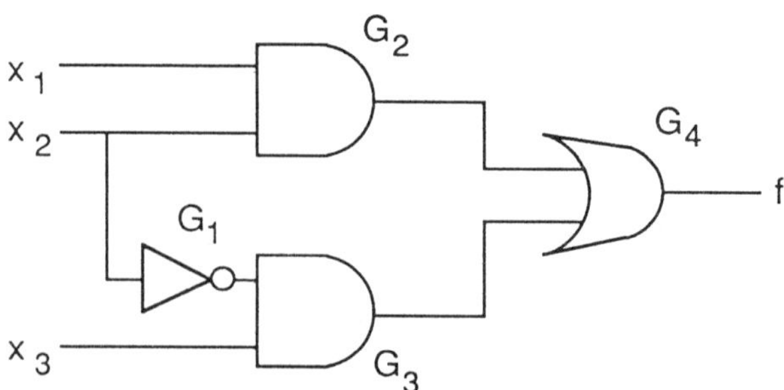

Fig. 3.1 An example circuit illustrating path sensitization

(s-a-0) fault on line x_3. We must first assign 1 to x_3 so that it has different logic values in the fault-free and faulty cases. Then we have to propagate the effect of the fault through gate G_3. This means that the output of the inverter G_1 must be set to 1. In order to further propagate the effect of the fault through G_4, we need to set the output of G_2 to 0. This process of propagating the effect of the fault to the circuit output is called forward drive. The next phase is called backward trace. In this phase signals are traced backwards to the primary inputs from each of the gates on the sensitized path so that the logic values specified in the forward drive phase can be realized. For example, in order to make the output of G_1 1, we must make $x_2 = 0$. Similarly, to make the output of G_2 0, at least one of its inputs must be 0. This has already been accomplished by making $x_2 = 0$. Thus the vector d01 is a test vector for the s-a-0 fault on x_3, where d denotes a "don't-care."

The problem with sensitizing a single path at a time is that we may not be able to derive a test vector for the fault even if the fault is detectable [SCHN67]. Thus in some cases we may need to sensitize more than one path at the same time. This is the basic idea behind D-algorithm, which is discussed next.

D-Algorithm

D-algorithm derives its name from the fact that the symbol D is used to denote the situation in which a line has a value 1 in the absence of the fault and 0 in its presence [ROTH66]. The complementary situation is denoted by $\overline{D}$. D-algorithm is a true algorithm in the sense that it can derive a test vector for any fault in the circuit, if such a vector exists. Otherwise, it declares that the fault is undetectable.

In order to understand how D-algorithm works we first need to understand the following terms: singular cover, primitive D-cube of a fault, propagation D-cubes and D-intersection.

Singular cover: The singular cover of a gate is simply a compact way of representing its truth table. For example, the singular cover of the two-input NAND gate is shown below.

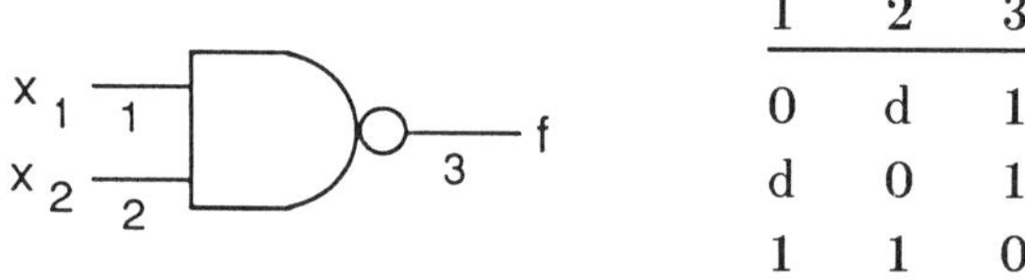

1	2	3
0	d	1
d	0	1
1	1	0

The rows in the singular cover are said to be cubes of the singular cover.

Primitive D-cube of a fault (pdcf): The primitive D-cube of a fault in a gate consists of the input vector which is required to produce a D or $\overline{D}$ at the output of the gate in the presence of the fault. For example, the primitive D-cubes for a s-a-0 fault on the output of the two-input NAND gate are

1	2	3
0	d	D
d	0	D

Propagation D-cubes: The propagation D-cubes of a gate are cubes which are required to propagate a D or $\overline{D}$ at one or more of its inputs to its output. The propagation D-cubes of the two-input NAND gate are

1	2	3
1	D	$\overline{D}$
D	1	$\overline{D}$
D	D	$\overline{D}$
1	$\overline{D}$	D
$\overline{D}$	1	D
$\overline{D}$	$\overline{D}$	D

D-intersection: The D-intersection of two cubes $A = (\alpha_1, \alpha_2, .., \alpha_n)$ and $B = (\beta_1, \beta_2, .., \beta_n)$, denoted by $A \cap B$, where the elements belong to the set $\{0, 1, d, D, \overline{D}\}$, is derived as follows:

(1)   If in some position i, $\alpha_i = \beta_i$, then the resultant cube also has the same value in that position.

(2)   If $\alpha_i = d$ ($\beta_i = d$), then the resultant cube has β_i (α_i) in position i.

(3)   If $\alpha_i \neq d$, $\beta_i \neq d$ and $\alpha_i \neq \beta_i$ at any position i, then the resultant cube is ϕ (null) in that position and $A \cap B = \Phi$, i.e. the null cube. In such a case we have an inconsistency and the D-intersection does not exist.

For example, let $A = (d0\overline{D}1d)$, $B = (D0\overline{D}1d)$ and $C = (00\overline{D}01)$. Then $A \cap B = (D0\overline{D}1d)$, $A \cap C = (00\overline{D}\phi1) = \Phi$ and $B \cap C = (\phi0\overline{D}\phi1) = \Phi$.

We next illustrate the above concepts through an example. Consider the circuit in Fig. 3.2. To derive test vectors for the faults in the circuit we first need to obtain the singular covers and

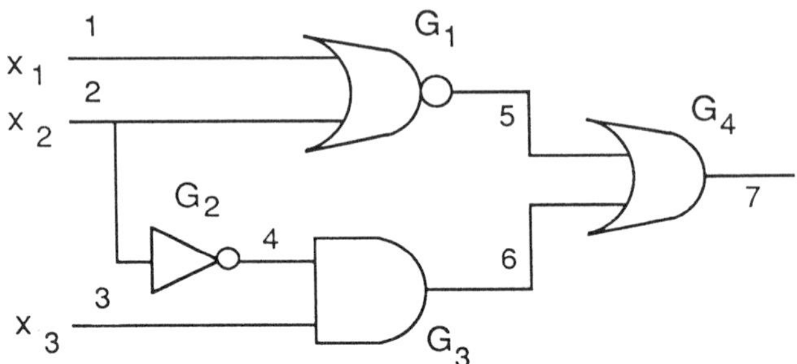

Fig. 3.2 An example circuit illustrating D-algorithm

propagation D-cubes for all the gates. These are shown in Tables 3.1 and 3.2. The blanks in these tables are treated as don't-care d. In Table 3.2 other propagation D-cubes can be obtained by

Table 3.1 Singular cover

Gate	Cube	1	2	3	4	5	6	7
G_1	a	d	1			0		
	b	1	d			0		
	c	0	0			1		
G_2	e		1		0			
	f		0		1			
G_3	g			d	0		0	
	h			0	d		0	
	i			1	1		1	
G_4	j					d	1	1
	k					1	d	1
	l					0	0	0

Table 3.2 Propagation D-cubes

Gate	Cube	1	2	3	4	5	6	7
G_1	m	D	0			$\overline{D}$		
	n	0	D			$\overline{D}$		
G_2	o		D		$\overline{D}$			
G_3	p			D	1		D	
	q			1	D		D	
G_4	r					D	0	D
	s					0	D	D

replacing D with $\overline{D}$ and $\overline{D}$ with D in each row. These have not been shown for the sake of brevity.

Suppose that we want to derive a test for a s-a-1 fault on line 5. To start the test generation process we first obtain the pdcf for this fault. This is shown as cube t in Table 3.3. At this point, line 5 has a $\overline{D}$ which can be propagated further. Thus this line is said to belong to the D-frontier. To propagate the effect through

Table 3.3 Test generation

Cube	1	2	3	4	5	6	7
t	1	0			$\overline{D}$		
$v = t \cap r'$	1	0			$\overline{D}$	0	$\overline{D}$
$w = v \cap g$	1	0		0	$\overline{D}$	0	$\overline{D}$
$x = w \cap e$	1	ϕ		0	$\overline{D}$	0	$\overline{D}$
$y = v \cap h$	1	0	0		$\overline{D}$	0	$\overline{D}$

gate G_4 we need to intersect cube t with the relevant propagation D-cube of gate G_4. This propagation D-cube can be derived from cube r by replacing D with $\overline{D}$ and is denoted by r$'$. The intersection of t and r$'$ is performed and is shown in the second row of Table 3.3. This results in a $\overline{D}$ at the circuit output. The above process is called D-drive and it is terminated when a D or $\overline{D}$ reaches the circuit output.

The next phase involves justifying the values derived in the D-drive phase and is called line justification. In order to obtain a 0 on line 6 we need to intersect cube v with a relevant cube of the singular cover of gate G_3. This is shown in row 3 of Table 3.3. Next we need to justify the value on line 4 by using the singular cover of gate G_2. However, this leads to an inconsistency. So we backtrack and justify a 0 on line 6 with the help of another cube of the singular cover of gate G_3. Thus the cube y is obtained. There is no inconsistency this time and the test vector is obtained as 100.

For large circuits, test generation may be very time-consuming due to the backtrackings. Therefore test generation is usually coupled with a process called fault simulation. This process determines which faults are detected by a given vector. Thus once a test vector is obtained for a given fault, it is fault simulated against other faults to determine which other faults are also detected. By using both test generation and fault simulation, the number of faults for which test vectors have to be generated can be considerably reduced.

For fault simulation we first need to determine the fault-free signal values in the circuit. Then the signal values have to be determined in the presence of the fault. The simplest approach is to use one fault at a time. However, it is also possible to simulate a number of faults at the same time by using a method called *parallel fault simulation* [SESH65]. The number of faults simulated usually depends on the word size of the host computer. It is possible to simulate an even larger number of faults by using a method called *deductive fault simulation* [ARMS72]. In this method the faults which will cause the signal values to be different from the fault-free values are deduced. This method, however, requires much more memory capacity.

3.2 BOOLEAN DIFFERENCE

The Boolean difference method is another approach that has been used for test generation [SELL68]. Consider a circuit which implements the function $f(X) = f(x_1, x_2, .., x_n)$ of n variables. We define

$$f_{\bar{x}_i} = f(x_1, x_2, \ldots, x_{i-1}, 0, x_{i+1}, \ldots, x_n)$$

$$f_{x_i} = f(x_1, x_2, .., x_{i-1}, 1, x_{i+1}, .., x_n).$$

The Boolean difference of the function $f(X)$ with respect to the variable x_i, denoted as $\dfrac{df(X)}{dx_i}$, is defined as follows:

$$\frac{df(X)}{dx_i} = f_{\bar{x}_i} \oplus f_{x_i}.$$

$\dfrac{df(X)}{dx_i} = 1$ if and only if $f_{\bar{x}_i} \neq f_{x_i}$. However, if $f_{\bar{x}_i} \neq f_{x_i}$, it means that changing the logic value at x_i also changes the output logic value. In other words, a change at x_i is observable at the output. Thus if we want to derive a test vector for a stuck-at fault on the line fed by x_i, we must have $\dfrac{df(X)}{dx_i} = 1$. The other condition which must be satisfied is that x_i must be assigned a logic value which is the complement of its stuck value. Therefore a test vector for a s-a-0 fault on this line must satisfy the following condition:

$$x_i \frac{df(X)}{dx_i} = 1.$$

We know that we must assign $x_i = 1$. Then to propagate the effect to the circuit output we must have $\frac{df(X)}{dx_i} = 1$. The above condition satisfies both the requirements simultaneously. Similarly, for detecting a s-a-1 fault on the line fed by x_i the test vector must satisfy

$$\overline{x}_i \frac{df(X)}{dx_i} = 1.$$

Example 3.1: Consider the circuit in Fig. 3.3. It implements the function $f(X) = f(x_1,x_2,x_3,x_4) = (x_1 + x_2)(x_3 + x_4)$. Suppose we are interested in detecting the s-a-1 fault on line x_3. We first derive the Boolean difference $\dfrac{df}{dx_3}$ as follows:

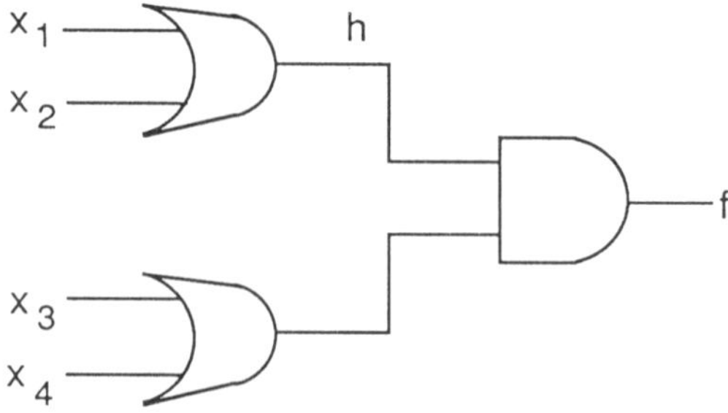

Fig. 3.3 An example circuit illustrating Boolean difference

$$\frac{df}{dx_3} = f_{\overline{x}_3} \oplus f_{x_3}$$

$$= (x_1 + x_2)x_4 \oplus (x_1 + x_2)$$

$$= x_1\overline{x}_4 + x_2\overline{x}_4.$$

Then from the condition $\overline{x}_3\dfrac{df}{dx_3} = 1$ we have

$$\overline{x}_3(x_1\overline{x}_4 + x_2\overline{x}_4) = 1$$

$$\text{or } (x_1\overline{x}_3\overline{x}_4 \mid x_2\overline{x}_3\overline{x}_4) = 1.$$

Therefore the test vectors are 1d00 and d100.

The Boolean difference approach can also be used to derive test vectors for faults which lie on the internal lines of the circuit. For example, suppose we want to detect the s-a-0 fault on line h in the circuit in Fig. 3.3. We first express the function f in terms of h and the other input variables as

$$f(X,h) = h(x_3 + x_4).$$

Then the Boolean difference is computed as

$$\frac{df(X,h)}{dh} = 0(x_3 + x_4) \oplus 1(x_3 + x_4)$$

$$= (x_3 + x_4)$$

The test vectors must satisfy the condition

$$h\frac{df(X,h)}{dh} = 1$$

$$\text{or } (x_1 + x_2)(x_3 + x_4) = 1$$

$$\text{or } x_1x_3 + x_1x_4 + x_2x_3 + x_2x_4 = 1.$$

Therefore the test vectors are 1d1d, 1dd1, d11d and d1d1. $\square$

3.3 FAULT COLLAPSING

Fault collapsing is a technique which is used to reduce the number of faults that need to be considered in order to completely test the circuit. It is easy to find many instances of two faults ϕ_1 and ϕ_2 which can not be distinguished by just observing the outputs of the circuit. For example, consider an OR gate with inputs $x_1, x_2, .., x_n$. If any one of these inputs or the output has a s-a-1 fault, the output will always be 1 for every input vector. Thus by just looking at the output it is not possible to find out which s-a-1 fault is present. Faults like these which can not be distinguished are said to be equivalent. Similarly, we can see that all the s-a-0 faults in an AND gate are equivalent. For a NOR (NAND) gate the set of equivalent faults consists of all the input s-a-1 (s-a-0) faults and the output s-a-0 (s-a-1) fault. Only one of the faults from the set of equivalent faults needs to be considered for test generation purposes.

Another way of reducing the number of faults that need consideration is through the use of the concept of fault dominance.

Let T_{ϕ_1} be the set of test vectors which detect fault ϕ_1 and T_{ϕ_2} be the set of vectors which detect fault ϕ_2. We say that fault ϕ_1 dominates fault ϕ_2 if $T_{\phi_2} \subseteq T_{\phi_1}$ [SCHE72]. Therefore if ϕ_1 dominates ϕ_2, any test vector which detects ϕ_2 also detects ϕ_1. For example, for an AND (NAND) gate a s-a-1 (s-a-0) fault at the output dominates a s-a-1 fault at any input. Similarly, for an OR (NOR) gate a s-a-0 (s-a-1) fault at the output dominates a s-a-0 fault at any input.

For an n-input gate there are $2(n + 1)$ single stuck-at faults that can occur at the inputs or the output. However, using fault equivalence and fault dominance, this can be collapsed to only $n + 1$ faults.

Using the concepts of fault equivalence and fault dominance for a circuit which consists of many gates, it is possible to show that a test set which detects single stuck-at faults at all the primary inputs and all the fanout branches also detects all single stuck-at faults in the circuit [TO73]. The primary inputs and fanout branches are called checkpoints.

Example 3.2: Consider the circuit in Fig. 3.4. This circuit has

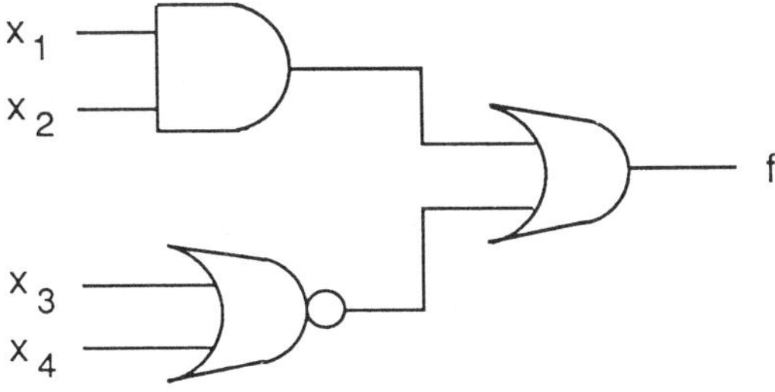

Fig. 3.4 An example circuit illustrating fault collapsing

no fanout. So the checkpoints are simply the four primary inputs. We need to consider both the s-a-0 an s-a-1 faults on each of these checkpoints for a total of eight faults. However, this fault list can be reduced further. We can see that the s-a-0 faults on lines x_1 and x_2 are equivalent. Similarly, the s-a-1 faults on lines x_3 and x_4 are equivalent. Therefore we need to consider one fault from each of the two pairs for a total of six faults only. If the fault collapsing techniques had not been used then we would have had to consider fourteen faults on the seven lines in the circuit. □

3.4 REDUNDANCY IN CIRCUITS

If there does not exist any vector which can detect some given fault ϕ then the circuit is said to be redundant with respect to ϕ. Consider the circuit in Fig. 3.5. This circuit is redundant with respect to the s-a-0 fault on line 3 since this fault does not change the function realized at f. Thus it is tempting to just ignore such faults. However, the presence of such faults can prevent one from detecting another fault which was detectable otherwise. For example, the s-a-1 fault on line 1 can be detected

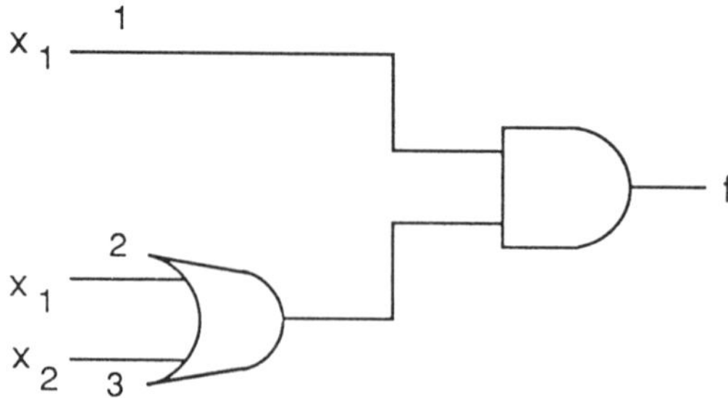

Fig. 3.5 A redundant circuit

by the vector 01. However, if a s-a-0 fault on line 3 is also present then the fault on line 1 can not be detected.

Another problem that arises for redundant circuits is that a lot of computational effort is wasted in trying to find a test vector for an undetectable fault. Therefore is it advisable to remove redundant lines and gates in the design phase itself. For example, the circuit in Fig. 3.5 implements the function $f = x_1(x_1 + x_2) = x_1$. Therefore all that is needed is a simple connection from x_1 to f. Unfortunately, for larger circuits identifying the redundant lines and gates can be a very time-consuming process. Therefore extra care needs to be taken while designing a circuit to avoid redundancies.

For redundant circuits, fanout stems of paths which reconverge later are also included in the set of checkpoints. However, it was pointed out in [ABRA86] that detecting all detectable single stuck-at faults at the primary inputs, fanout stems and fanout branches in a redundant circuit does not guarantee the detection of all detectable single stuck-at faults in the circuit. Test generation for some additional lines may be needed. However, in the rest of the chapter we will assume that the circuit under test is irredundant.

3.5　TESTING OF DOMINO CMOS CIRCUITS

Domino CMOS [KRAM82] has become a popular technique in the last few years. We discussed its advantages and limitations in Chapters 1 and 2. In this section we will see how various faults can be detected in domino CMOS circuits. Single fault detection in domino CMOS circuits was first considered in [OKLO84,

MANT84].

3.5.1 Testing of Gates with Series-Parallel Network

It is quite easy to generate tests for a domino CMOS circuit
from its gate-level model. Consider the domino CMOS circuit in
Fig. 3.6 in which the logic network of the complex gate is series-
parallel in nature. Its gate-level model is shown in Fig. 3.7. We
first derive a test set for all single stuck-at faults in the gate-level
model. In the model in Fig. 3.7 the checkpoints are the four primary inputs. There are eight possible single stuck-at faults that
can occur on these lines. Out of these faults the s-a-1 faults on
lines 1 and 2 are equivalent and the the s-a-1 faults on lines 3 and
4 are equivalent. Therefore we only need to derive test vectors for
the remaining six faults in order to detect all the single stuck-at

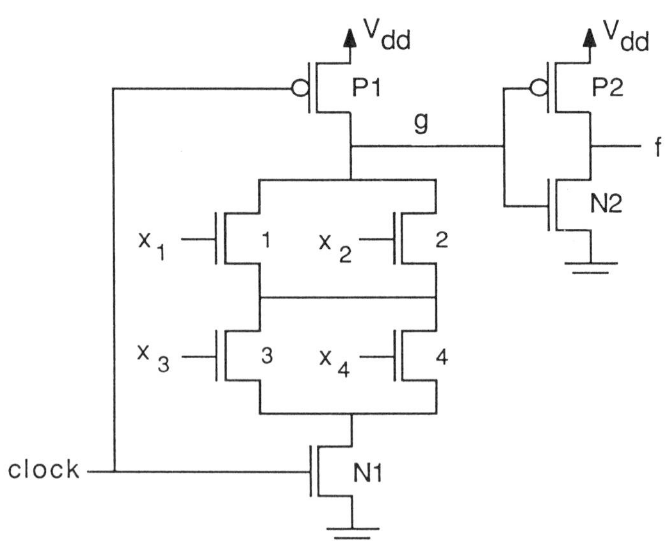

Fig. 3.6 A domino CMOS circuit

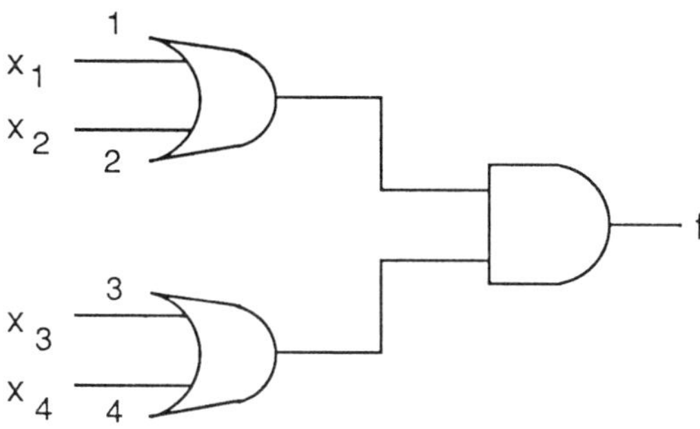

Fig. 3.7 The gate-level model

faults in the circuit. Any algorithm which is applicable to gate-level circuits can be used, e.g. D-algorithm or Boolean difference. One of the possible test sets that can be derived is {0011, 0101, 1100, 1010}.

Let us see how the above test set can be used for detecting faults in the domino CMOS circuit in Fig. 3.6. Suppose that the vectors are applied to this circuit in the sequence shown. These vectors and the resultant outputs are shown in Table 3.4.

Table 3.4 Test set

x_1	x_2	x_3	x_4	f
0	0	1	1	0
0	1	0	1	1
1	1	0	0	0
1	0	1	0	1

We first look at faults in the nMOS logic network consisting of transistors 1, 2, 3 and 4. Testing of these faults is carried out during the evaluation phase when clock = 1. Consider the first vector 0011. This tests for the s-a-1 faults on lines 1 and 2 in the gate-level model in Fig. 3.7. Correspondingly, it detects the stuck-on faults in transistors 1 and 2 in the domino CMOS circuit. If either of these two transistors is stuck-on, a conduction path is activated from node g to ground resulting in a logic 0 at g. However, in the fault-free case node g should maintain its precharged value of logic 1. So we see that the test for a s-a-1 fault on the line fed by $x_1(x_2)$ in the gate-level model is also a test for a stuck-on fault in the transistor fed by $x_1(x_2)$ in the domino CMOS circuit.

Next consider the vector 0101. This tests for the s-a-0 faults on lines 2 and 4 in the gate-level model. In the domino CMOS circuit this vector would activate a unique conduction path from node g to ground through transistors 2 and 4 in the fault-free case. Therefore if either transistor 2 or 4 is stuck-open, this path can not be activated and the fault will be detected. We note that to detect these stuck-open faults we do not require two-pattern tests. This is due to the fact that the initialization of node g is already taken care of in the precharge phase.

Similarly, the vector 1100 can be shown to detect the stuck-on fault in transistor 3 or 4 and 1010 can be shown to detect the stuck-open fault in transistor 1 or 3. We can therefore see that corresponding to a stuck-on (stuck-open) fault in the nMOS logic network there is a s-a-1 (s-a-0) fault in the gate-level model. Thus a test set which detects all single stuck-at faults in the gate-level model also detects all the stuck-on and stuck-open faults in the nMOS logic network. This result holds even if the network has a non-series-parallel structure, as we will see later. However, for such structures one has to be careful in deriving the gate-level model.

Let us now look at the faults which are not in the nMOS logic network. A stuck-open fault in transistor N1 will result in node g behaving in a s-a-1 fashion because it can be precharged but can not discharge. Thus there will be an equivalent s-a-0 fault at f. Similarly, a stuck-open fault in transistor P2 will also result in a s-a-0 fault at f. This will be detected by vectors 0101 and 1010.

The remaining transistors in which stuck-open faults have to be considered are N2 and P1. For a stuck-open fault in N2, initialization is not automatically taken care of. We have to first make f = 1 with an initialization vector and then try to pull it down to 0 with a test vector. When the fault is present, f will remain at 1 and the fault will be detected. This means that we need a two-pattern test which causes a $1 \rightarrow 0$ transition at f. In Table 3.4 this is done by the two-pattern test $<0101, 1100>$. A $1 \rightarrow 0$ transition at f implies a $0 \rightarrow 1$ transition at g in Fig. 3.6. Therefore the same two-pattern test also detects a stuck-open fault in transistor P1. One may argue that if P1 becomes stuck-open then g can not get precharged and whatever previous charge was stored at that node would be lost through stray capacitances. Therefore g would behave in a s-a-0 fashion, which would require a single test vector. Nonetheless, the point is that whenever we detect N2 stuck-open, we also automatically detect P1 stuck-open.

We now look at stuck-on faults in N1, P1, N2 and P2. Consider the P1 stuck-on fault first. When any vector which makes g = 0 is applied, e.g. 0101 or 1010, there is conduction between g and ground in the evaluation phase. At the same time there is conduction between V_{dd} and g because of the stuck-on fault. Thus a large amount of current will be drawn by the circuit and the fault will be detected if current is monitored. However, if only logic values are monitored, the fault will be detected only if the voltage at g maps to logic 1. Similarly, a stuck-on fault in transistor P2 (N2) can be detected by any test vector which results in 0(1) at f in the fault-free case, if current monitoring is done.

Detecting a stuck-on fault in transistor N1 is slightly tricky. Suppose a vector is applied which results in 0 at g in the fault-free case, for example, vector 0101 or 1010. Even when N1 is stuck-on there is a conduction path from g to ground during the evaluation phase. Then when the precharge phase starts there is a conduction path from V_{dd} to ground for some time before the vector is changed. If this time interval is large enough so that there is a measurable increase in the current drawn by the circuit, then the fault will be detected. Furthermore, if the next vector that arrives is such that it also makes g logic 0, then the process of current monitoring may be aided further.

If only logic monitoring is done, the notion of p-dominance and n-dominance can be used to determine which stuck-on faults in the clocked transistors or the inverter are detectable. A CMOS gate is said to be p-dominant (n-dominant) if its output is 1 (0) when both its pMOS and nMOS networks conduct due to a fault [OKLO84]. A CMOS gate can be made p-dominant or n-dominant by adjusting the resistances of its two networks. For example, in Fig. 3.6 suppose we want to make the complex gate p-dominant and the inverter n-dominant. We could increase the width of transistor P1 slightly so that its resistance becomes much less than the resistance of the nMOS network. Thus when there is conduction between V_{dd} and ground due to a stuck-on fault in P1 or N1, g will remain at 1. Therefore P1 stuck-on will be detectable by logic monitoring whereas N1 stuck-on will not be. Similarly, if the inverter is made n-dominant then N2 stuck-on will be detectable, whereas P2 stuck-on will not be.

We finally consider the single stuck-at faults in the domino CMOS circuit in Fig. 3.6. A s-a-1 (s-a-0) fault at the input of any transistor in the nMOS logic network is equivalent to a stuck-on (stuck-open) fault which has already been covered before. The stuck-at faults on nodes g and f are also detected by the test set in

Table 3.4. Thus the only line that remains to be considered is the stem of the clock line. Suppose this line has a s-a-0 fault. This results in an equivalent double fault consisting of N1 stuck-open and P1 stuck-on. This in turn is equivalent to a s-a-1 fault on g. Therefore this fault is detected by any test vector that makes g = 0. Next consider a s-a-1 fault on the stem of the clock line. The corresponding equivalent double fault is N1 stuck-on and P1 stuck-open. Since node g can not be precharged, we can argue, as before, that whatever charge it had stored is lost through stray capacitances. The node g could also lose its charge if a vector is applied to the circuit which makes g = 0. Thus node g would behave in a s-a-0 fashion. This can be detected by any test vector which makes g = 1 in the fault-free case.

From the above discussions we see that a test set derived for all the single stuck-at faults in the gate-level model can be used to detect the stuck-open, stuck-on and stuck-at faults in the domino CMOS circuit. The only restriction is that there should be at least one 1→0 transition at the output of the inverter.

3.5.2 Testing of Gates with Non-Series-Parallel Network

Consider a domino CMOS circuit with non-series-parallel nMOS logic network as shown in Fig. 3.8. Its gate-level model can be derived using a technique similar to the one given in [REDD84] and is shown in Fig. 3.9. The gate-level modeling technique given in [REDD84] will be discussed in more detail in Chapter 4 (Section 4.1.1.2). The test set which detects all single stuck-at faults in Fig. 3.9 is given in Table 3.5. This test set has the required 1→0 transition at f and hence will also detect the single faults in the domino CMOS circuit in Fig. 3.8. However, if we had tried to derive the test set directly from the domino CMOS circuit, we

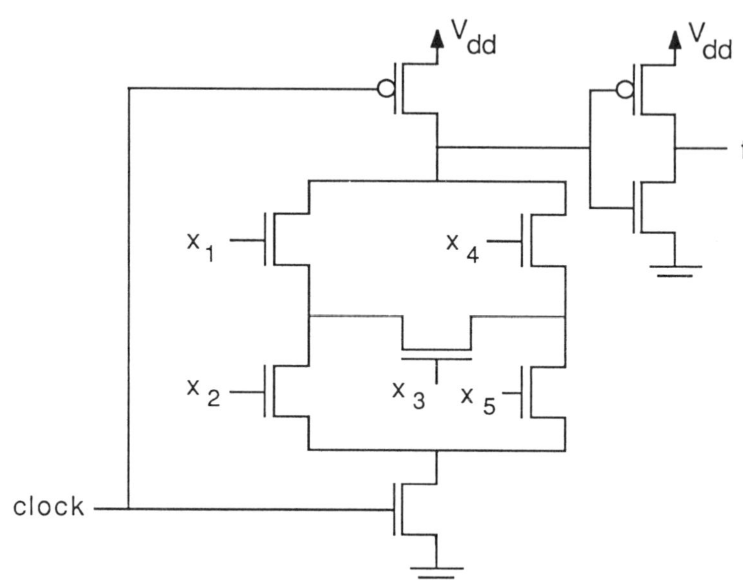

Fig. 3.8 A domino CMOS circuit

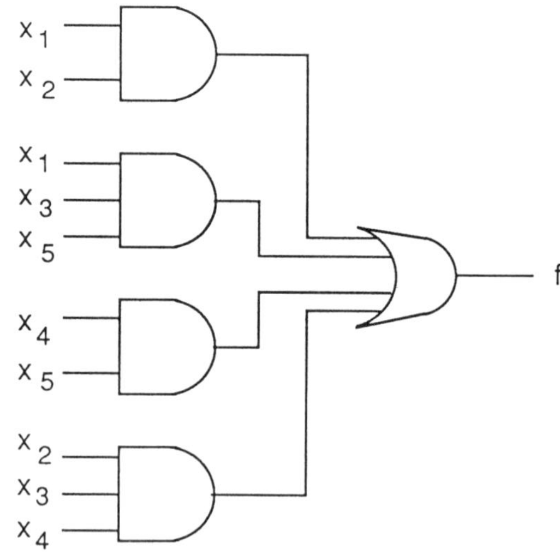

Fig. 3.9 The gate-level model

Table 3.5 A test set for the gate-level model

x_1	x_2	x_3	x_4	x_5	f
1	1	0	0	0	1
1	0	1	0	1	1
0	0	0	1	1	1
0	1	1	1	0	1
1	0	1	1	0	0
0	1	1	0	1	0
1	0	0	0	1	0
0	1	0	1	0	0

could have come up with the following test set: {10101, 01110, 10001, 01010}, with the tests applied in the sequence shown. This test set also detects all the single faults. It is a subset of the test set shown in Table 3.5. Therefore a test set derived from the gate-level model of a domino CMOS circuit with a non-series-parallel nMOS logic network is sufficient, but may not be necessary, for detecting all single faults in the domino CMOS circuit. However, not many real-life CMOS circuits contain non-series-parallel networks, the reason being that it is not always easy to obtain such networks from the functional description.

We must emphasize once again that it is important to derive the gate-level model for such networks by using a technique similar to the one given in [REDD84] to ensure that the test set derived from the gate-level model also suffices for the domino CMOS circuit.

3.5.3 Testing of a General Circuit

We next consider a more complex domino CMOS circuit given in Fig. 3.10. The gate-level model for this circuit is shown in

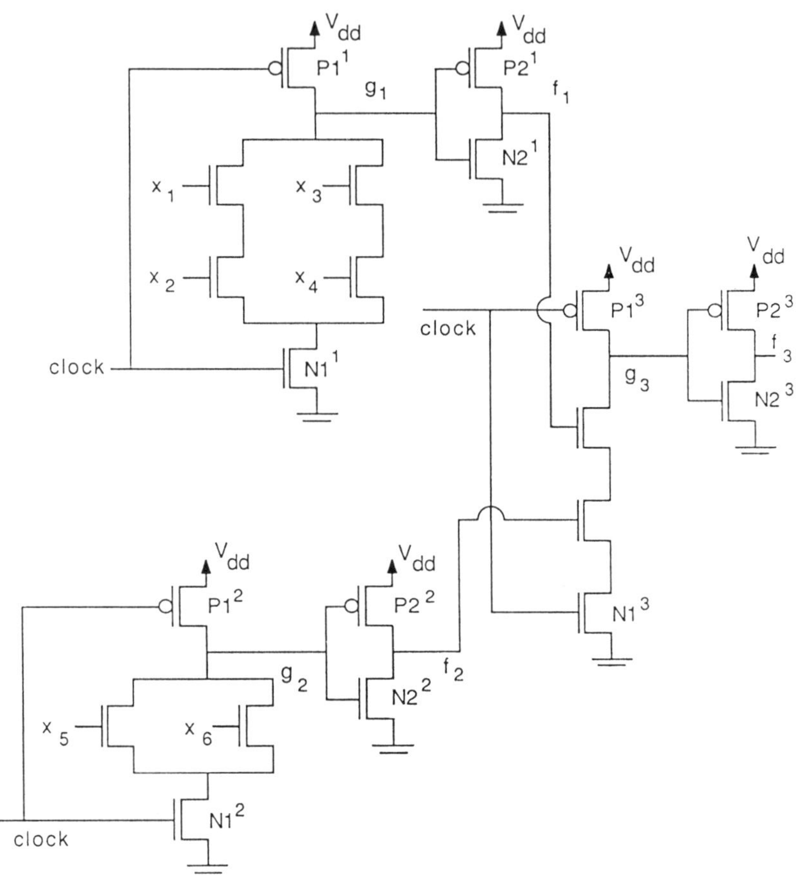

Fig. 3.10 A domino CMOS circuit

Fig. 3.11. A stuck-at fault test set for this gate-level model can be
obtained by any of the standard methods mentioned before. A
test set which detects all the single stuck-at faults in this model is
given in Table 3.6. Let us see how this test set can be used for the

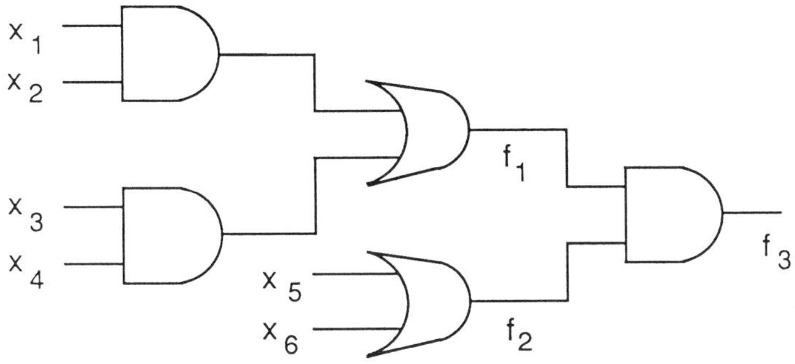

Fig. 3.11 The gate-level model

Table 3.6 A test set for the circuit in Fig. 3.11

x_1	x_2	x_3	x_4	x_5	x_6	f_1	f_2	f_3
1	1	0	0	0	1	1	1	1
1	0	1	0	0	1	0	1	0
0	0	1	1	1	0	1	1	1
0	1	0	1	1	0	0	1	0
1	1	0	0	0	0	1	0	0

domino CMOS circuit. We know that we have to feed each of the inverters with a two-pattern test so that we can detect the stuck-open faults in transistors $N2^1$, $N2^2$ and $N2^3$. This two-pattern test should result in a $1 \rightarrow 0$ transition at the output of the inverter in the fault-free case. Also, the test vector of the two-pattern test should sensitize a path from the inverter output to the circuit output. In Table 3.6 the two-pattern test $<110001, 101001>$ consisting of the first two vectors satisfies the above conditions for nodes f_1 and f_3. Similarly, the two-pattern test $<010110, 110000>$ consisting of the last two vectors satisfies the above condition for node f_2. It is interesting to note that if a two-pattern test is such that both its initialization and test vector sensitize the same path to the circuit output, then it tests for a stuck-open fault in the nMOS transistor of every inverter along this path.

From earlier discussions we know that the other single stuck-open and stuck-on faults in the circuit will also get detected by the test set in Table 3.6, if both logic and current monitoring are done. Similarly, the stuck-at faults at the inputs of the transistors in the different nMOS logic networks or the inputs or outputs of the different inverters will also be detected. The stuck-at faults on the clock lines, however, deserve special attention. The clock lines which are fed to the different domino CMOS gates are in fact all connected together, although not explicitly shown as such in Fig. 3.10. Therefore a stuck-at fault on this line may affect more than one domino CMOS gate. For example, let the clock line be shorted to ground resulting in a s-a-0 fault. This will affect all the domino CMOS gates. In particular, transistors $N1^1$, $N1^2$ and $N1^3$ will be stuck-open, and transistors $P1^1$, $P1^2$ and $P1^3$ will be stuck-on. We know that this will result in a s-a-0 fault on lines f_1, f_2 and f_3. Let us look at the corresponding lines in the gate-level model. When more than one line in the circuit is stuck at the same value the circuit is said to have a unidirectional stuck-at fault. Therefore in the gate-level model we have a unidirectional s-a-0 fault on lines f_1, f_2 and f_3. This fault is detected by any vector which detects the individual f_1 s-a-0, f_2 s-a-0 or f_3 s-a-0 faults. For example, both the vectors 110001 and 001110 will detect the unidirectional fault. We may wonder if this is a happy coincidence or if this result can be generalized. Fortunately, it turns out that it can be generalized.

In general, consider a gate-level model in which $F = (f_1, f_2, .., f_n)$ consists of lines which correspond to the outputs of the inverters in the domino CMOS circuit that is being modeled. Let T^0 be a set of tests which detects the single s-a-0 faults on these lines. Let $T_i^0 \in T^0$ be a test for a s-a-0 fault on line $f_i \in F$. Then the claim is that T_i^0 also detects all the unidirectional s-a-0 faults on lines from F in which a s-a-0 fault on line f_i is included. The reason is that if we look towards the circuit output in the gate-level model from any of the lines in F we will not encounter any inversions. Thus the test T_i^0 which detects a s-a-0 fault on line f_i can not be invalidated by s-a-0 faults on other lines in F. These other faults can only further aid in testing. Another way of looking at this is to note that T_i^0 must result in a 1 at f_i in the fault-free case and it must also sensitize a path to the circuit output. Since there are no inversions along the sensitized path the circuit output must also have a 1 in the fault-free case. When f_i is s-a-0 the circuit output also changes to 0. When one or more other lines from F are also s-a-0, the circuit output can never become 1 due to the multiple fault. Therefore the multiple fault will also be detected. Similarly, we can show that a test which detects a s-a-1 fault on line f_i also detects all unidirectional s-a-1 faults on two or more lines from F, if the unidirectional fault includes the s-a-1 fault on f_i.

With the above explanation we can better understand how the stuck-at faults on the clock lines are detected. We considered the s-a-0 faults on these lines before. We next consider a s-a-1 fault. Suppose that the clock line is shorted to V_{dd}. This will affect all the three domino CMOS gates in Fig. 3.10. The equivalent faults will be $N1^1$, $N1^2$ and $N1^3$ stuck-on, and $P1^1$, $P1^2$ and $P1^3$ stuck-open. Since nodes g_1, g_2 and g_3 can no longer be precharged, they will lose their charge and behave in a s-a-0 fashion. These nodes can also lose their charge if a vector activates a conduction path from them to ground. Therefore there will be a unidirectional s-a-1 fault on lines f_1, f_2 and f_3. The vectors from Table 3.6 which will detect this fault are 101001, 010110 and 110000. Thus the stuck-at faults on the clock line are automatically detected. In the examples above, we assumed that the clock line is shorted to ground or V_{dd}, which affects all the domino CMOS gates in the circuit. However, we should keep in mind that the stuck-at fault model is just that - a model for physical failures that actually occur in the circuit. Therefore a stuck-at fault on the clock line does not necessarily imply that the line is shorted to ground or V_{dd}. Thus it is possible that not all branches of the clock line will be affected by the stuck-at fault. However, from previous explanations we know that the fault will still be detected. Therefore a test set derived for single stuck-at faults in the gate-level model, if properly ordered, also detects the single stuck-open, stuck-on and stuck-at faults in the domino CMOS circuit.

Finally, let us consider a bridging fault in the domino CMOS circuit in Fig. 3.10. Suppose that lines f_1 and f_2 are bridged together. Any vector which tries to feed (0,1) or (1,0) to (f_1, f_2) will detect this fault if current monitoring is done. The reason is that there will be a low-resistance conduction path from V_{dd} to ground in the presence of the bridging fault. The following vectors from Table 3.6 detect this fault: 101001, 010110, 110000. If only the logic value is monitored at the circuit output f_3 then this fault may not be detected. Similarly, a bridging fault between two inputs feeding an nMOS logic network at the first level in the circuit in Fig. 3.10 will also be detected by current monitoring if they are fed complementary values. Of course, we assume that the primary inputs of this circuit are fed from other CMOS circuits.

3.5.4 Ordering of Tests

For a small circuit the ordering of tests so that every inverter receives its two-pattern test is quite straightforward. However, for a large circuit, ensuring the proper ordering may require some effort. Alternative ways of taking care of this problem have also been suggested. One of the suggested methods is simply to apply the test set two times to the circuit [WUND86]. Consider a domino CMOS circuit in which the nMOS transistor $N2^i$ in some inverter with output f_i is stuck-open. If the circuit is irredundant then there must exist one or more vectors in the test set which make $f_i = 1$. After such a vector is applied, node f_i should behave in a s-a-1 fashion for some time because of the stuck-open fault in $N2^i$. Thus when the test set is applied the second time, the fault is expected to be detected. Now consider the stuck-open fault in the clocked transistor $P1^i$ in the domino CMOS gate associated with this inverter. We have previously assumed that since the node g_i (output of the domino CMOS gate) can not be precharged, it will lose its charge and behave in a s-a-0 fashion. This assumption is strengthened by applying the vector(s) mentioned above. A vector which makes $f_i = 1$ makes $g_i = 0$. Therefore any residual charge at g_i will definitely be lost. Again the equivalent fault at f_i will be a s-a-1 fault which can be expected to be detected when the test set is applied the second time.

Another way of avoiding the need for ordering tests has been suggested in [JHA88]. This method involves the application of a set of initialization vectors to the domino CMOS circuit before the application of the test set derived from the gate-level model. There is no need to order the vectors in the test set. The set of initialization vectors is such that, given any inverter in the domino CMOS circuit, there is at least one vector in the set which makes the output of the inverter logic 1. For example, for the circuit in Fig. 3.10 the initialization set consists of only one vector 111111. The vectors from the test set shown in Table 3.6 can be applied in any order after this initialization vector is applied, as shown in Table 3.7. If the initialization vector had not been applied and the

Table 3.7 An alternate test set

x_1	x_2	x_3	x_4	x_5	x_6	f_1	f_2	f_3
1	1	1	1	1	1	1	1	1
1	0	1	0	0	1	0	1	0
0	1	0	1	1	0	0	1	0
1	1	0	0	0	0	1	0	0
1	1	0	0	0	1	1	1	1
0	0	1	1	1	0	1	1	1

test set had been applied in the order shown in Table 3.7 then the single stuck-open faults in transistors $N2^1$ and $N2^3$ would not be detected. However, because of the initialization vector, the corresponding inverter output will get initialized to 1 when either of these two stuck-open faults is present. Thus the fault will be detected when the test set is applied.

The basic assumption behind this technique is that the output of the faulty inverter behaves in a s-a-1 fashion throughout the application of the test set. To ensure that this assumption remains valid even for large test sets, the test set can be divided into two or more parts and the initialization set can be applied at the beginning of each part. In [JHA88] multiple fault detection in domino CMOS circuits is also considered.

Robust testability. A test set derived for a domino CMOS circuit is inherently robust against invalidation due to circuit delays and timing skews, as explained in Chapter 2. If an SCVS circuit with feedback pMOS transistors is used (see Chapter 2) instead of the domino CMOS circuit then the test set will also be robust against invalidation due to charge sharing [BARZ84].

Other circuits. Many other variations of domino CMOS

circuits have also been presented in literature [GONC83, FRIE84, LEE86, PRET86, HWAN89]. A design for testability technique has been presented in [LING87] for NORA CMOS circuits [GONC83]. For the circuits given in [FRIE84, LEE86, PRET86, HWAN89], although no formal testing techniques have been presented in literature yet, the technique described in this section should prove to be a useful starting point.

3.6 TESTING OF CVS CIRCUITS

The cascode voltage switch (CVS) logic technique [HELL84] is an extension of the domino CMOS technique. We have discussed how a DCVS and an SCVS circuit works in Chapter 1. A CVS circuit has the potential for being faster and more area-efficient than static CMOS circuits [HELL84]. It also overcomes the limitation of domino CMOS circuits, which can only implement non-inverting functions. However, not much has been done in the area of CVS circuit testing. The first attempts in this direction were made in [MONT85, BARZ85]. Recently, testability of particular kinds of DCVS circuits has been considered in [JIIA89, JIIA90].

Consider the DCVS EX-OR gate in Fig. 3.12. The feedback pMOS transistors which may be needed to alleviate the problem of charge sharing (see Chapter 2) have not been shown in Fig. 3.12 for the sake of simplicity. A test set for this circuit that has been derived directly from the transistor diagram is shown in Table 3.8 [JHA89]. In this table Φ_1 denotes the set of stuck-open faults that are detected by a single vector, Φ_2 denotes the set of stuck-on faults in the nMOS logic network which are detected by logic monitoring and Φ_3 denotes the set of stuck-on faults in the circuit detected by current monitoring. For example, the vector 100 detects stuck-open faults in transistors 1, 8, 10, 11 and 16 at

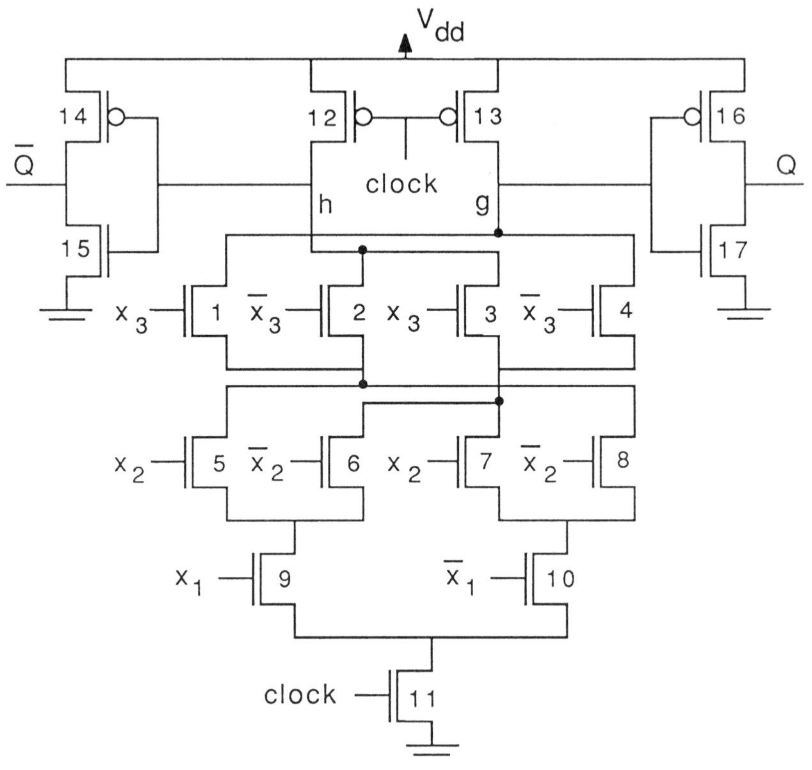

Fig. 3.12 A DCVS EX-OR gate

Table 3.8 Test set for a 3-input EX-OR circuit

Test $x_3x_2x_1$	Q	Φ_1	at	Φ_2	at	Φ_3
1 0 0	1	1,8,10,11,16	Q	2,7,9	$\overline{Q}$	13,14,17
0 1 1	0	2,5,9,11,14	$\overline{Q}$	1,6,10	Q	12,15,16
0 0 1	1	4,6,9,11,16	Q	3,5,10	$\overline{Q}$	13,14,17
1 1 0	0	3,7,10,11,14	$\overline{Q}$	4,8,9	Q	12,15,16

output Q, stuck-on faults in transistors 2, 7 and 9 at output $\overline{Q}$,
and stuck-on faults in transistors 13, 14 and 17 using current

monitoring. The stuck-open faults in transistors 13 and 17 are detected by the two-pattern test $<100, 011>$. This two-pattern test creates the necessary $1\rightarrow0$ transition at Q in the fault-free case. One can argue, as before, that a stuck-open fault in transistor 13 will result in a s-a-0 fault at g and hence a s-a-1 fault at Q. Nevertheless, the fault will still be detected by the test vector 011 of the two-pattern test. Similarly, the stuck-open faults in transistors 12 and 15 are detected by the two-pattern test $<011, 001>$. This creates a $0\rightarrow1$ transition at Q and hence a $1\rightarrow0$ transition at $\overline{Q}$ in the fault-free case. Testing of the stuck-on fault in transistor 11 is tricky as in the domino CMOS circuit case. However, the same arguments can be applied again for the detection of this fault. Similarly, the stuck-at faults in this circuit can also be shown to be detected by the test set in Table 3.8.

The nMOS logic network of the circuit in Fig. 3.12 is non-series-parallel in nature. We know that for such a network a test set derived from the gate-level model may be sufficient but not necessary. In other words, it may contain tests which are not needed. A DCVS complex gate is usually non-series-parallel in nature because the nMOS logic required for realizing Q and the nMOS logic required for realizing $\overline{Q}$ frequently share many transistors. Therefore test generation from the gate-level model may be inappropriate for DCVS circuits. Although no efficient formal technique for generating a deterministic test set for arbitrary DCVS circuits consisting of any number of DCVS gates has appeared yet, the graph-theoretic approach employed in [CHIA82, CHIA83] for MOS networks may hold promise. Similarly, transistor-level test generation algorithms such as those given in [AGRA84, REDD85, SHIH86], which have been devised for MOS circuits, may also be adapted to DCVS circuits.

The high testability of some particular DCVS circuits as shown in [JHA89, JHA90] is very encouraging. For example, in

[JHA89] it is shown that any DCVS parity tree (a tree of DCVS EX-OR gates) can be tested for all detectable single stuck-open, stuck-on and stuck-at faults by at most five tests. In some cases only four tests are necessary. This result holds irrespective of the number of inputs to the parity tree and the size of the constituent EX-OR gates. If only one output is desired from the parity tree instead of two complementary outputs, then the size of the test set becomes eight. The size of the robust test set for static CMOS parity trees on the other hand increases with the depth of the tree [KUND88].

For a DCVS ripple-carry adder the size of the test set is four irrespective of the number of stages in it [JHA90]. For a k-input DCVS one-count generator (a circuit which generates a count of the number of 1s in its inputs) the size of the test set is at most $5\lceil \log_2(k+1) \rceil - 5$ [JHA90]. For some particular values of k, the size of the test set is as small as four. A static CMOS implementation of these circuits would require many more tests. An additional advantage of CVS circuits over static CMOS circuits is that their test sets are always robust (see Chapter 2). Therefore in addition to possible area and speed advantages over static CMOS, CVS circuits also have a large testability advantage.

REFERENCES

[ABRA86] M. Abramovici, P. R. Menon, and D. T. Miller, "Checkpoint faults are not sufficient faults for test generation," *IEEE Trans. Comput.*, vol. C-35, pp. 769-771, Aug. 1986.

[AGRA84] P. Agrawal and S. M. Reddy, "Test generation at MOS level," in *Proc. Int. Conf. Computers, Systems & Signal Proc.*, Bangalore, India, Dec. 1984.

[ARMS66] D. B. Armstrong, "On finding a nearly minimal set of fault detection tests for combinational logic nets," *IEEE Trans. Electron. Comput.*, pp. 66-73, Feb. 1966.

[ARMS72] D. B. Armstrong, "A deductive method for simulating faults in logic circuits," *IEEE Trans. Comput.*, vol. C-21, pp. 464-471, May 1972.

[BARZ84] Z. Barzilai et al., "Accurate fault modeling and efficient simulation of SCVS circuits," in *Proc. Int. Conf. Computer Design*, Port Chester, NY, pp. 42-47, Oct. 1984.

[BARZ85] Z. Barzilai et al., "Accurate fault modeling and efficient simulation of differential CVS circuits," in *Proc. Int. Test Conf.*, Philadelphia, PA, pp. 722-729, Oct. 1985.

[CHIA82] K. W. Chiang and Z. G. Vranesic, "Test generation for MOS complex gate networks," in *Proc. Int. Symp. Fault-Tolerant Comput.*, Santa Monica, CA, pp. 149-157, June 1982.

[CHIA83] K. W. Chiang and Z. G. Vranesic, "On fault detection in CMOS logic networks," in *Proc. Design Automation Conf.*, Miami Beach, FL, pp. 50-56, June 1983.

[FRIE84] V. Friedman and S. Liu, "Dynamic logic CMOS circuits," *IEEE J. Solid-State Circuits*, vol. SC-19, no. 2, pp. 263-266, Apr. 1984.

[FUJI83] H. Fujiwara and T. Shimono, "On the acceleration of test generation algorithms," in *Proc. Int. Symp. Fault-Tolerant Comput.*, Milan, Italy, pp. 98-105, June 1983.

[GOEL81] P. Goel, "An implicit enumeration algorithm to generate tests for combinational logic circuits," *IEEE Trans. Comput.*, vol. C-30, pp. 215-222, Mar. 1981.

[GONC83] N. F. Goncalves and H. J. De Man, "NORA: A racefree dynamic CMOS technique for pipelined logic structures," *IEEE J. Solid-State Circuits*, vol. SC-18, no. 3, pp. 261-266, June 1983.

[HELL84] L. G. Heller et al., "Cascode voltage switch logic: A differential CMOS logic family," in *Proc. Int. Solid-State Circuits Conf.*, pp. 16-17, Feb. 1984.

[HWAN89] I. S. Hwang and A. L. Fisher, "Ultrafast compact 32-bit CMOS adders in multiple-output domino logic," *IEEE J. Solid-State Circuits*, vol. 24, no. 2, pp. 358-369, Apr. 1989.

[JHA88] N. K. Jha, "Testing for multiple faults in domino-CMOS logic circuits," *IEEE Trans. CAD*, vol. 7, pp. 109-116, Jan. 1988.

[JHA89] N. K. Jha, "Fault detection in CVS parity trees: Application to SSC CVS parity and two-rail checkers," in *Proc. Int. Symp. Fault-Tolerant Comput.*, Chicago, IL, pp. 407-414, June 1989.

[JHA90] N. K. Jha, "Testing of differential cascode voltage switch one-count generators," *IEEE J. Solid-State Circuits*, (in press).

[KRAM82] R. H. Krambeck, C. M. Lee, and H.-F. S. Law, "High-speed compact circuits with CMOS," *IEEE J. Solid-State Circuits*, vol. SC-17, no. 3, pp. 614-619, June 1982.

[KUND88] S. Kundu and S. M. Reddy, "Robust tests for parity trees," in *Proc. Int. Test Conf.*, Washington, D.C., pp. 680-687, Sept. 1988.

[LEE86] C. M. Lee and E. W. Szeto, "Zipper CMOS," *IEEE Circuits & Devices*, pp. 10-17, May 1986.

[LING87] N. Ling and M. A. Bayoumi, "An efficient technique to improve NORA CMOS testing," *IEEE Trans. Circuits & Systems*, vol. CAS-34, no. 12, pp. 1609-1611, Dec. 1987.

[MANT84] S. R. Manthani and S. M. Reddy, "On CMOS totally self-checking circuits," in *Proc. Int. Test Conf.*, Philadelphia, PA, pp. 866-877, Oct. 1984.

[MONT85] B. Montoye, "Testing scheme for differential cascode voltage switch circuits," *IBM Tech. Disclosure Bulletin*, vol. 27, no. 10B, pp. 6148-6152, Mar. 1985.

[OKLO84] V. G. Oklobdzija and P. G. Kovijanic, "On testability of CMOS-domino logic," in *Proc. Int. Symp. Fault-Tolerant Comput.*, Orlando, FL, pp. 50-55, June 1984.

[PRET86] J. A. Pretorius, A. S. Shubat, and C. A. T. Salama, "Latched domino CMOS logic," *IEEE J. Solid-State Circuits*, vol. SC-21, no. 4, pp. 514-522, Aug. 1986.

[REDD84] S. M. Reddy, V. D. Agrawal, and S. K. Jain, "A gate-level model for CMOS combinational logic circuits with application to fault detection," in *Proc. Design Automation Conf.*, Albuquerque, NM, pp. 504-509, June 1984.

[REDD85] M. K. Reddy, S. M. Reddy, and P. Agrawal, "Transistor level test generation for MOS circuits," in *Proc. Design Automation Conf.*, Las Vegas, pp. 825-828, June 1985.

[ROTH66] J. P. Roth, "Diagnosis of automata failures: A calculus and a method," *IBM J. Res. & Dev.*, vol. 10, pp. 278-291, July 1966.

[SCHE72] D. R. Schertz and G. Metze, "A new representation for faults in combinational digital circuits," *IEEE Trans. Comput.*, vol. C-21, pp. 858-866, Aug. 1972.

[SCHN67] P. R. Schneider, "On the necessity to examine D-chains in diagnostic test generation - an example," *IBM J. Res. & Dev.*, vol. 11, p. 114, Jan. 1967.

[SCHU88] M. H. Schultz, E. Trischler, and T. M. Sarfert, "SOCRATES: A highly efficient ATPG system," *IEEE Trans. CAD*, vol. 7, pp. 126-137, Jan. 1988.

[SELL68] F. F. Sellers, M. Y. Hsiao, and C. L. Bearnson, "Analyzing errors with the Boolean difference," *IEEE Trans. Comput.*, vol. C-17, pp. 676-683, July 1968.

[SESH65] S. Seshu, "On an improved diagnosis program," *IEEE Trans. Electron. Comput.*, vol. EC-14, pp. 76-79, Feb. 1965.

[SHIH86] H.-C. Shih and J. A. Abraham, "Transistor-level test generation for physical failures in CMOS circuits," in *Proc. Design Automation Conf.*, Las Vegas, pp. 243-249, June-July

1986.

[TO73]   K. To, "Fault folding for irredundant and redundant combinational circuits," *IEEE Trans. Comput.*, vol. C-22, pp. 1008-1015, Nov. 1973.

[WUND86] H.-J. Wunderlich and W. Rosenstiel, "On fault modeling for dynamic MOS circuits," in *Proc. Design Automation Conf.*, Las Vegas, pp. 540-546, June-July 1986.

ADDITIONAL READING

[HA85]   D. S. Ha and S. M. Reddy, "On the design of testable domino PLAs," in *Proc. Int. Test Conf.*, Philadelphia, PA, pp. 567-573, Oct. 1985.

[OKLO86] V. G. Oklobdzija and R. K. Montoye, "Design-performance trade-offs in CMOS-domino logic," *IEEE J. Solid-State Circuits*, vol. SC-21, no. 2, pp. 304-306, Apr. 1986.

[PRET86] J. A. Pretorius, A. S. Shubat, and C. A. T. Salama, "Charge distribution and noise margins in domino CMOS logic," *IEEE Trans. Circuits & Systems*, vol. CAS-33, no. 8, pp. 786-793, Aug. 1986.

PROBLEMS

3.1.   For the circuit in Fig. 3.13, find a test for the s-a-0 fault on line h using path sensitization.

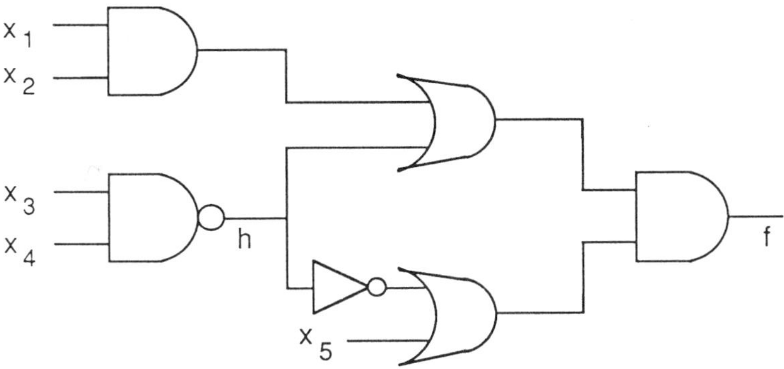

Fig. 3.13

3.2. For a 3-input OR gate, derive the singular cover, the set of propagation D-cubes and the set of primitive D-cubes for all the single stuck-at faults.

3.3. Use D-algorithm to find a test for the fault x_3 s-a-0 in Fig. 3.13. What are the other faults detected by this test?

3.4. Find all tests for the fault x_1 s-a-1 in the circuit in Fig. 3.14 using the Boolean difference technique.

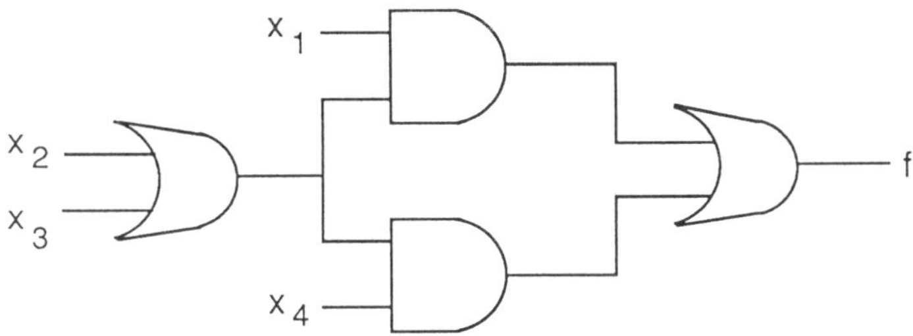

Fig. 3.14

3.5. What are the checkpoints in the circuit in Fig. 3.14? Find a minimal set of single stuck-at faults whose detection will result in the detection of all single stuck-at faults in the circuit.

3.6. Are all the single stuck-at faults in the circuit in Fig. 3.15 detectable? If not, does the presence of any undetectable stuck-at fault prevent the detection of another detectable fault?

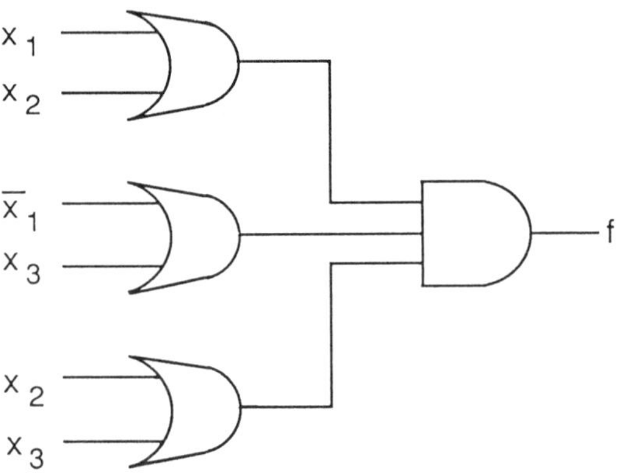

Fig. 3.15

3.7. In a domino CMOS circuit why is every complex gate always followed by an inverter?

3.8. For the domino CMOS circuit in Fig. 3.16 find a test vector for the following single faults:

(**a**) a stuck-open fault in transistor 1 and N2,

(**b**) a stuck-on fault in transistor 3,

(**c**) a s-a-0 fault on the clock line.

3.9. What are the faults in the circuit in Fig. 3.16 which require current monitoring? If the complex gate is designed to be p-dominant and the inverter is designed to be n-dominant, which of these faults can be guaranteed to be detected by only monitoring the logic values at output f.

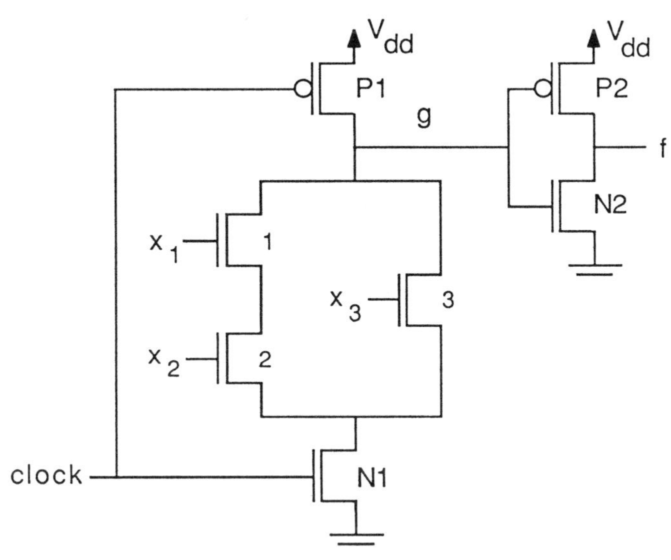

Fig. 3.16

3.10. Consider the domino CMOS circuit given in Fig. 3.10 and its gate-level model given in Fig. 3.11. Corresponding to the $N1^2$ stuck-open fault in this domino CMOS circuit, what is the equivalent fault in the gate-level model? What is a test vector for this fault?

3.11. Design a domino CMOS circuit to implement the function $f = x_1(x_2 + x_3) + x_2x_4$. Derive a test set for your circuit which will detect all the single stuck-open, stuck-on and stuck-at faults. Assume that both logic and current monitoring are done.

3.12. Design a 2-input SCVS EX-OR gate. Derive a test set for detecting all the single stuck-open, stuck-on and stuck-at faults in it.

3.13. For the DCVS parity tree given in Fig. 3.17 derive a minimal test set which detects all the single stuck-open, stuck-on and stuck-at faults. (Hint: a test set consisting of only four vectors can be found).

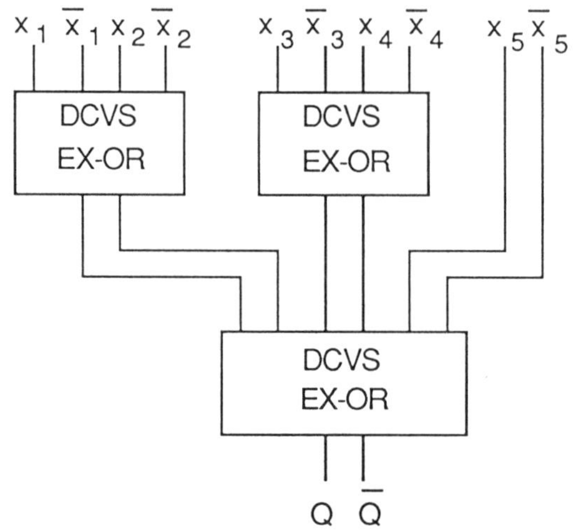

Fig. 3.17

3.14. Suppose only the output Q is desired in the circuit in Fig. 3.17. Then the modified circuit will be as shown in Fig. 3.18 with the final EX-OR gate being SCVS instead of DCVS. For this parity tree derive a minimal test set for detecting the single stuck-open, stuck-on and stuck-at faults.

3.15. Design a 3-input DCVS carry circuit whose Q output is 1 if and only if two or more of its inputs are 1. Use as few transistors for implementing the nMOS logic network as possible. Derive a test set for this circuit which detects all the single stuck-open, stuck-on and stuck-at faults.

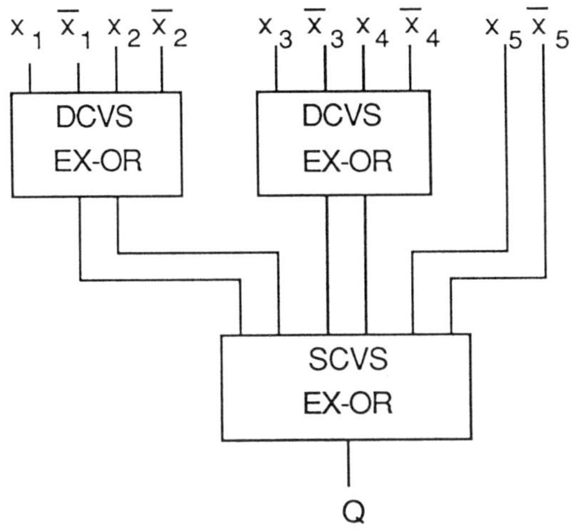

Fig. 3.18

3.16. Assume that a DCVS full-adder consists of the 3-input
DCVS EX-OR gate shown in Fig. 3.12 and the DCVS carry
circuit that was designed in Prob. 3.15. Derive a minimal
test set which detects all the single stuck-open, stuck-on and
stuck-at faults in the full-adder.

Chapter 4
TEST GENERATION FOR STATIC CMOS CIRCUITS

In the previous chapter we discussed testing methods for combinational dynamic CMOS circuits. In this chapter we will concentrate on combinational static CMOS circuits. Test generation for a static CMOS circuit, as in the case of a dynamic CMOS circuit, can be done from either its gate-level model or its transistor-level description. We will discuss both the approaches.

From Chapter 2 we know that a test set for a static CMOS circuit, unless carefully derived, can be invalidated by circuit delays. To avoid test invalidation we need robust test sets. However, to ensure robustness of the test set we need to pay the price of increased test generation time. Some testing engineers may wish to sacrifice some fault coverage rather than pay this price. Therefore non-robust test generation is also important. We will discuss both non-robust and robust test generation in this chapter.

It is worth pointing out here that there are some static CMOS circuits for which complete robust testing is not possible [REDD83]. These circuits can be redesigned for testability. Many methods are known which can generate robustly testable designs. They will be discussed in Chapter 6.

We will first discuss some non-robust test generation methods for combinational static CMOS circuits based on either the gate-level model or the transistor-level description.

4.1 NON-ROBUST TEST GENERATION

The fault models that are generally used for static CMOS circuits are the stuck-open, stuck-on and stuck-at fault models. A stuck-open fault transforms a combinational logic circuit into a sequential circuit with 1-bit memory having two states. One is the faulty state, the other is the fault-free one. In the theory of sequential circuits a vector sequence that drives a sequential machine to a known state regardless of its initial state is called a *synchronizing sequence*. If a k-state machine has one or more synchronizing sequences, then the length of at least one of the sequences is at most $\dfrac{k\,(k+1)(k-1)}{6}$ [KOHA78]. In case of single stuck-open faults, $k = 2$. Substituting this value of k in the above expression yields the value 1. Therefore the length of the synchronizing sequence is no more than 1. Thus the initialization of the circuit for a single stuck-open fault, if initialization is at all possible, can be achieved with a single pattern. A second pattern is needed to test for the fault. Thus a two-pattern test is sufficient to detect a detectable stuck-open fault. In the above explanation we have not taken circuit delays into account. For robust testing, multi-pattern tests may be necessary [REDD86].

After two-pattern tests are derived, some post-processing is required to derive an efficient test set. For example, suppose that the two-pattern tests $<X,Y>$, $<X,Z>$ and $<Y,Z>$ have been derived for three stuck-open faults. One can trivially derive the sequence $<X,Y,X,Z,Y,Z>$ from these two-pattern tests. However, another sequence which also includes all the two-pattern tests, but requires only five vectors, is $<X,Y,Z,X,Z>$. The objective of a testing engineer is to apply as small a sequence as possible to reduce the time it takes to test the circuit.

Many methods have been presented in literature for generating a test set, which are not necessarily robust, for detecting single stuck-open faults. We will first discuss some of the methods which use the gate-level model of the CMOS circuit for test generation.

4.1.1 Test Generation from a Gate-Level Model

Notable among the methods which use the gate-level model are the Jain-Agrawal method [JAIN83, JAIN85], the Reddy-Agrawal-Jain method [REDD84a] and the Chandramouli method [CHAN83] In the first method a modeling block called the B-block is used to represent the memory state caused by the stuck-open fault. In the second method such a block is not used. In the third method stuck-open fault tests are derived by properly arranging the stuck-at fault tests. These methods are presented next.

4.1.1.1 The Jain-Agrawal Method

Jain and Agrawal use AND, OR, NOT gates and the B-block to model CMOS circuits. A gate-level model is first derived for both the nMOS and pMOS networks of a CMOS gate in the circuit. A series (parallel) connection of transistors is replaced with an AND (OR) gate. The inputs to the gate-level model of the pMOS network are complemented. This is done to take care of the fact that a pMOS transistor conducts when its input is 0. Let the output of the gate-level model which models the pMOS (nMOS) network be S_1 (S_0). S_1 and S_0 are fed to the B-block. The truth table of the B-block is shown in Table 4.1.

Table 4.1 Truth table of the B-block

S_1	S_0	Y
0	0	M
0	1	0
1	0	1
1	1	0

When $S_1 = 0$ and $S_0 = 1$, it means that there is a conduction path through the nMOS network but not the pMOS network. Therefore the output Y of the B-block is 0. Similarly, for $S_1 = 1$ and $S_0 = 0$, since only the pMOS network conducts, Y is 1. When both the networks conduct due to a fault, $S_1 = S_0 = 1$. In such a case it is assumed that the nMOS network dominates over the pMOS network. In Chapter 3 we referred to this phenomenon as n-dominance. If this assumption is not true for some particular case, then Y can be set to "unknown" for $S_1 = S_0 = 1$. Finally, $S_1 = S_0 = 0$ implies that both the networks are non-conducting. This situation can arise when a stuck-open fault is present. In such a case the output of the faulty gate remains in the floating state, and the logic value at the output depends on what the previous logic value was. This introduces memory and is denoted by M.

Example 4.1: Consider the static CMOS complex gate in Fig. 4.1. Its logic model is shown in Fig. 4.2. $\square$

When a CMOS circuit consists of more than one CMOS gate, the logic model for each gate in the circuit is derived and these models are interconnected together in the same way that the modeled CMOS gates are interconnected.

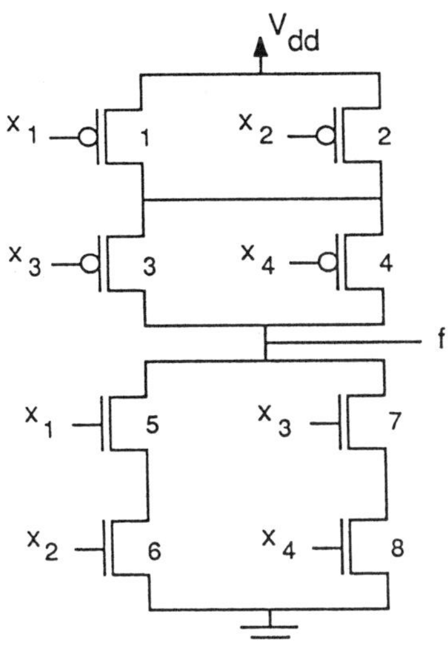

Fig. 4.1 A static CMOS complex gate

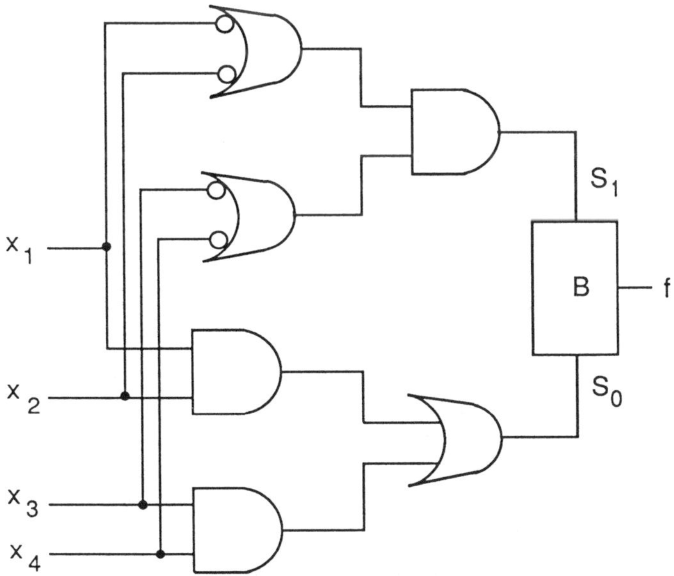

Fig. 4.2 The logic model for the CMOS gate in Fig. 4.1

The stuck-open, stuck-on and stuck-at faults in a CMOS circuit are translated into stuck-at faults in the logic model. To see how this translation is done let us refer to the NAND gate and its logic model shown in Fig. 4.3. The relationship among faults in the CMOS NAND gate and the faults in the logic model is shown in Table 4.2.

Since only stuck-at faults need to be considered in the logic

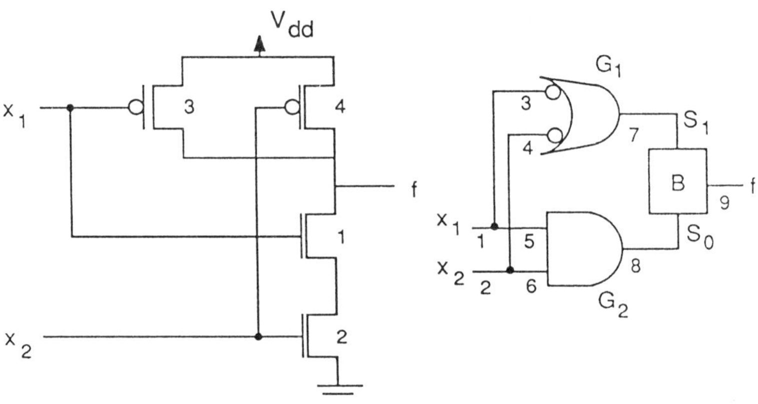

Fig. 4.3 A NAND gate and its logic model

Table 4.2 Equivalent faults

Fault in circuit	Fault in model
x_1 stuck-at 1(0)	line 1 stuck-at 1(0)
x_2 stuck-at 1(0)	line 2 stuck-at 1(0)
f stuck-at 1(0)	line 9 stuck-at 1(0)
transistor 1 on(open)	line 5 stuck-at 1(0)
transistor 2 on(open)	line 6 stuck-at 1(0)
transistor 3 on(open)	line 3 stuck-at 0(1)
transistor 4 on(open)	line 4 stuck-at 0(1)

model we can use the basic D-algorithm [ROTH66], as described in Chapter 3, to derive tests for the faults. In order to apply this algorithm we need to derive the singular cover and propagation D-cubes for the B-block that is used in the logic model. These are shown in Tables 4.3 and 4.4 respectively. In Table 4.3 d denotes a don't-care. In Table 4.4 the first four cubes do not involve the memory state M. However, the value of M needs to be specified for the remaining propagation D-cubes. Consider the seventh cube which can be derived by combining the following rows of the truth table.

Table 4.3 Singular cover

S_1	S_0	Output
0	0	M
d	1	0
1	0	1

Table 4.4 Propagation D-cubes

S_1	S_0	Output
1	D	$\overline{D}$
1	$\overline{D}$	D
D	$\overline{D}$	D
$\overline{D}$	D	$\overline{D}$
0	D	$\overline{D}(1)$
0	$\overline{D}$	$D(1)$
D	0	$D(0)$
$\overline{D}$	0	$\overline{D}(0)$
D	D	$\overline{D}(1)$
$\overline{D}$	$\overline{D}$	$D(1)$

S_1	S_0	Output
1	0	1
0	0	M

If the previous state M of the output was 1 then it would remain 1 irrespective of the value on S_1. Thus the D on S_1 will not be propagated to the output. However, if the previous state M was 0 then the D on S_1 can be propagated to the output as D. Thus the value in the parenthesis in Table 4.4 denotes the value that M must have for a D or $\overline{D}$ to propagate to the output.

For stuck-open faults we first need to initialize the circuit. The initialization algorithm is derived from D-algorithm with the essential difference that D-drive is not used and a modified singular cover is used for the faulty gate. Initialization requires that both the faulty and fault-free circuits behave identically. Therefore if a line is stuck-at 0(1) then 1(0) should not be applied to it during initialization. All singular cubes which attempt to assign a value, which is the complement of the stuck value, to the line can thus be deleted from the singular cover.

A second copy of the circuit is used for initialization so that the circuit on which D-algorithm is being run is not disturbed. All lines are first assigned don't-cares. The required initialization logic value of 0 or 1 is then placed on the target line. Line justification and consistency check procedures are repeatedly used until the initialization vector is obtained, using backtracking if necessary.

The following example illustrates the test generation procedure.

Example 4.2: Suppose we want to derive a test for the stuck-open fault in transistor 3 in the NAND gate of Fig. 4.3. The corresponding fault in the logic model is line 3 stuck-at 1 (see Table 4.2). The primitive D-cube of the fault (pdcf) (see Chapter 3 for the definition of pdcf) will result in lines 3 and 4 (and hence lines 1 and 2) being assigned 0 and 1 respectively. This generates a $\overline{D}$ on line 3 which is propagated to line 7 as a D. With $x_1 = 0$ and $x_2 = 1$, the resultant value on line 8 is 0. From Table 4.4 we know that $S_1 = D$ and $S_2 = 0$ will propagate to the output as D if the output is initialized to 0. In the initialization step all lines are first assigned a don't-care d. Then 0 is placed on line 9. We know that for initialization both the faulty and fault-free circuits should behave identically. In the faulty circuit, line 7 has 0. Thus S_1 is made 0 for the initialization step as well. From the second singular cube in the singular cover of the B-block given in Table 4.3 we know that S_0 or line 8 has to be made 1 in order to obtain a 0 on line 9. From the singular cover of gate G_2 we then obtain $x_1 = x_2 = 1$. Thus a two-pattern test for the stuck-open fault in transistor 3 is $<11,01>$. $\square$

4.1.1.2 The Reddy-Agrawal-Jain Method

Reddy, Agrawal and Jain presented a simpler gate-level model in [REDD84a] for static CMOS circuits for application to fault detection than the one given by Jain and Agrawal discussed in the previous section. The Reddy-Agrawal-Jain model does not need a memory block such as the B-block. Furthermore, only the pMOS network or the nMOS network is modeled, but not both. In addition, a non-series-parallel network can also be modeled by this method.

We next present the method which lets us derive the equivalent gate-level model from the pMOS networks of the CMOS gates in the circuit. This method can be trivially modified to be applicable to the nMOS networks. The method has the following two steps:

Step 1: Apply procedure REDUCE (given ahead) to the pMOS network of each gate in the given CMOS circuit.

Step 2: Apply procedure EQUIVALENT (also given ahead) to the reduced networks in Step 1.

Procedure REDUCE

(1)   Associate a unique index with each transistor in the pMOS network.

(2)   Replace each series or parallel connection of pMOS transistors with a single pMOS transistor. Associate a set of indices with this pMOS transistor which is the union of the indices of the replaced pMOS transistors.

(3)   Repeat Step 2 until no further reduction is possible.

(4)   Label all nodes of the reduced network by integers and the inputs of the pMOS transistors by letters.

(5)   Find every loop-free conduction path from V_{dd} to the output node and express it as a product of the labels of the pMOS transistors lying on it.

(6)   Derive the gate function G_f as a sum-of-products expression in which the products are from Step 5.

Procedure EQUIVALENT

(1)   For every pMOS transistor in the reduced network, whose input is say Z, derive an equivalent circuit of the part of the pMOS network represented by the index set of Z as follows: replace a series (parallel) connection of transistors by AND (OR) gates and complement the inputs.

(2)   By employing AND gates and an OR gate, combine the outputs of the circuits derived in Step 1 to realize the gate function G_f.

Example 4.3: Consider the CMOS complex gate shown in Fig. 4.4 in which the two networks are non-series-parallel in nature. We will first derive the equivalent gate-level model from its pMOS network. The reduced network that can be obtained from this network is shown in Fig. 4.5. The set of all loop-free conduction paths from V_{dd} to f is {124, 1234, 1324, 134}. Thus $G_f = AD + ACE + BCD + BE$. A, B, C, D and E can be derived as shown in Fig. 4.6 using procedure EQUIVALENT. The final gate-level model is shown in Fig. 4.7.

As mentioned earlier, it is also possible to derive the gate-level model from the nMOS network. For example, the equivalent circuits for the nMOS transistors in the reduced nMOS network can be derived as shown in Fig. 4.8. Note that there is a direct connection between x_3 and I, and x_8 and J. There is no need to complement the inputs since an nMOS transistor conducts when its input is 1. The equivalent gate-level model is shown in Fig. 4.9. In this model an inverter is placed at the output because when there is conduction through the nMOS network, the output is 0, not 1. □

Either of the two models in Fig. 4.7 or Fig. 4.9 can be used for test generation. Let C be a CMOS circuit and E_C be its gate-

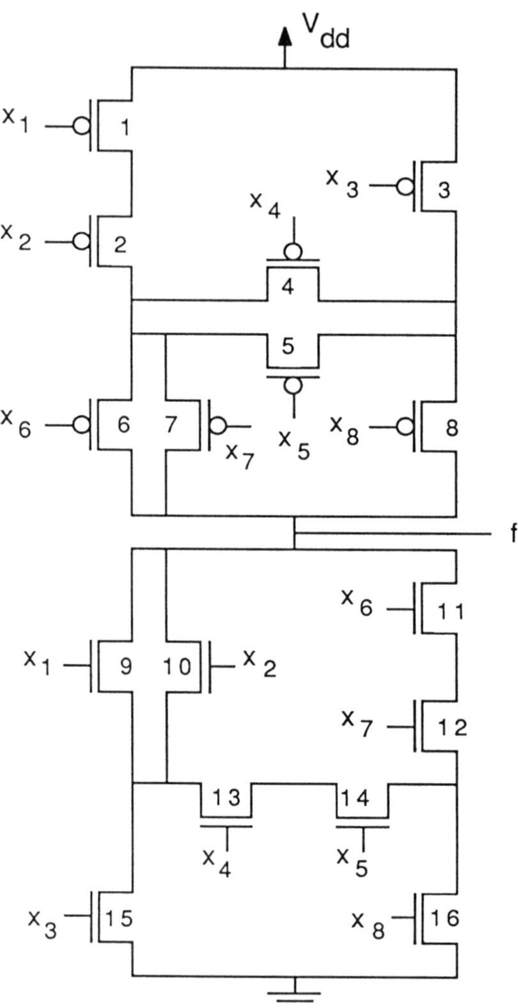

Fig. 4.4. A CMOS complex gate

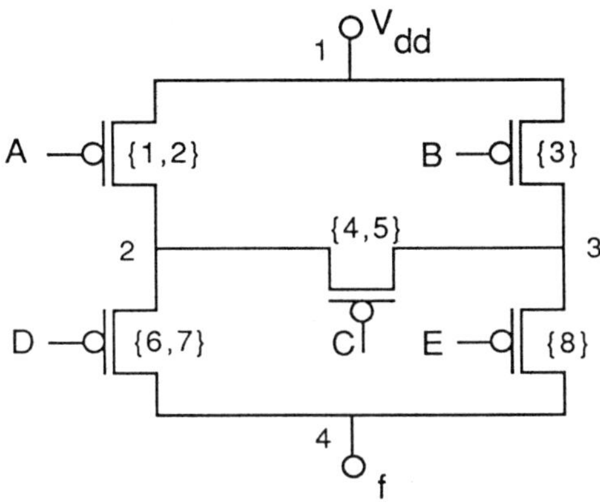

Fig. 4.5 The reduced network

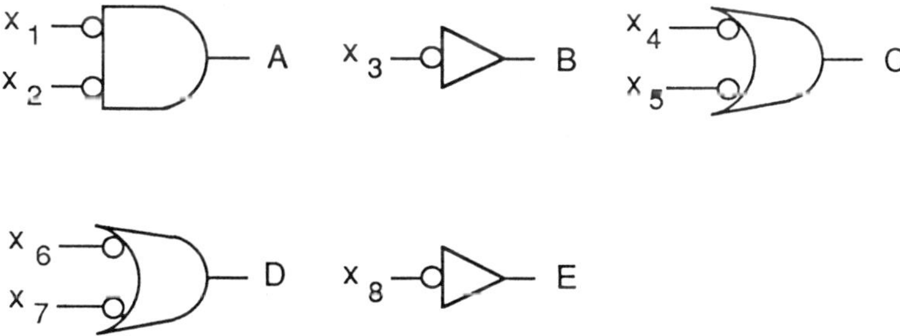

Fig. 4.6 Equivalent circuits for pMOS transistors
in Fig. 4.5

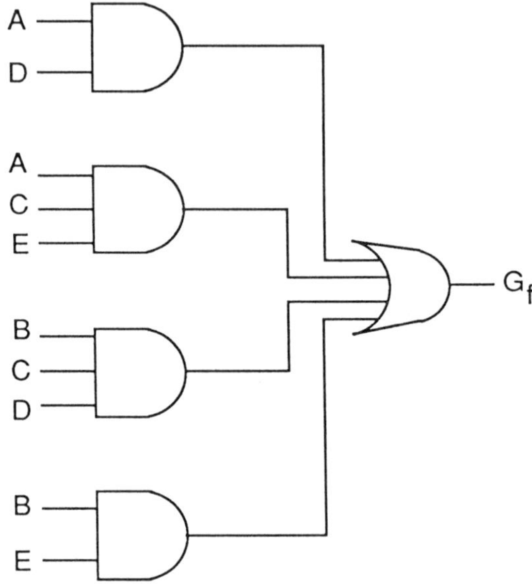

Fig. 4.7 The gate-level model

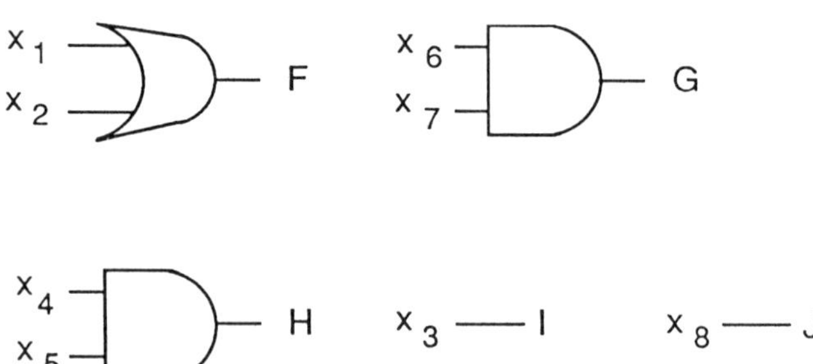

Fig. 4.8 Equivalent circuits for nMOS transistors
in the reduced nMOS network

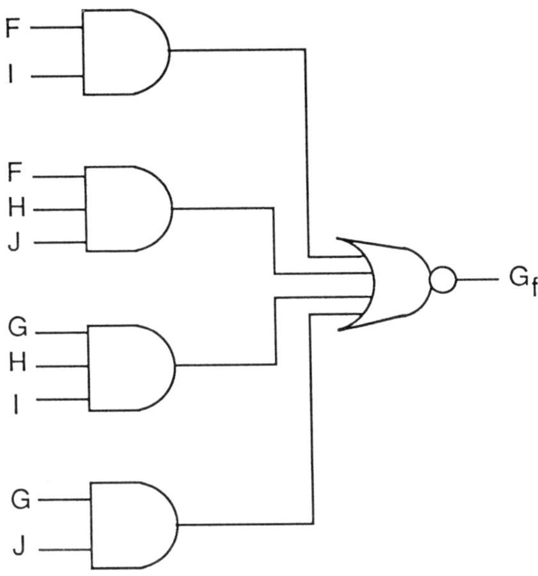

Fig. 4.9 Equivalent gate-level model

level model. The following result has been proved in [REDD84a]:

Result 4.1: A test vector to detect a stuck-at 1(0) fault on line m in E_C is a test vector to detect a stuck-at 1(0) fault on the line corresponding to m in C, stuck-open fault in the pMOS (nMOS) transistor and stuck-on fault in the nMOS (pMOS) transistor, whose input is connected to the line in C which corresponds to m.

Of course, for a stuck-open fault an initialization vector will have to be applied first, and for a stuck-on fault current monitoring may be necessary. Any gate-level test generation algorithm can be used to derive the tests for the stuck-at faults in the gate-level model.

Example 4.4: Suppose we want to derive a two-pattern test for the stuck-open fault in the nMOS transistor labeled 9 in the complex gate in Fig. 4.4. Let us use the gate-level model defined by the circuits in Figs. 4.6 and 4.7. According to Result 4.1, we first need to derive a test for the line x_1 stuck-at 0 in the gate-level model. From Figs. 4.6 and 4.7 we find that $x_1 = 1$, $x_2 = 0$, $B = C = E = 0$, $D = 1$ provides us with one possible test vector. One of the vectors that this translates to is $(x_1, x_2, x_3, x_4, x_5, x_6, x_7, x_8) = 101110d1$. The initialization vector should make $f = 1$. One such vector is $001110d1$, which also happens to be a test vector for the stuck-open fault in transistor 1, which is the pMOS transistor fed by x_1. $\square$

We therefore see that even though the equivalent gate-level model in Fig. 4.7 was derived from the pMOS network, it can still be used to derive tests for the faults in the nMOS network. The gate-level model in Fig. 4.9 can be similarly used for test generation for faults in both the networks.

4.1.1.3 The Chandramouli Method

Chandramouli showed that a test set which detects all stuck-open faults in an irredundant static CMOS circuit can be constructed by rearranging the stuck-at fault tests derived for the circuit [CHAN83]. He mentioned that the checkpoints for stuck-open fault testing are the set of primary inputs which do not fan out and the set of all fanout branches. In other words, if all the stuck-open faults in transistors connected to the above checkpoints in an irredundant CMOS circuit are detected by a test set then the test set detects all stuck-open faults in the circuit. Note that for stuck-at fault testing, primary input fanout stems are also included as checkpoints (see Section 3.3). However, these fanout stems need not be included for stuck-open faults as only the direct input of the transistor (fanout branch) is subject to the stuck-open fault, not the fanout stem.

This method is best illustrated through an example.

Example 4.5: Consider a two-level CMOS circuit whose gate-level model is as shown in Fig. 4.10. The checkpoints in this circuit are the six primary inputs. Let the test for a stuck-at 0(1) fault on checkpoint i be T_i^0 (T_i^1). Consider the checkpoint 1 fed by x_1. We can easily derive T_1^0 to be 111000 and T_1^1 to be 011000. The two-pattern test $<T_1^0, T_1^1>$ detects the stuck-open fault in the pMOS transistor fed by x_1 in the CMOS circuit that is being modeled. Similarly, $<T_1^1, T_1^0>$ detects the stuck-open fault in the nMOS transistor fed by x_1. We can combine the two two-pattern tests into a three-pattern test $<T_1^0, T_1^1, T_1^0>$. This three-pattern test detects stuck-open faults in both the transistors fed by x_1. Similar three-pattern tests can be derived for every checkpoint in the circuit and a test set can be derived from them. However, we

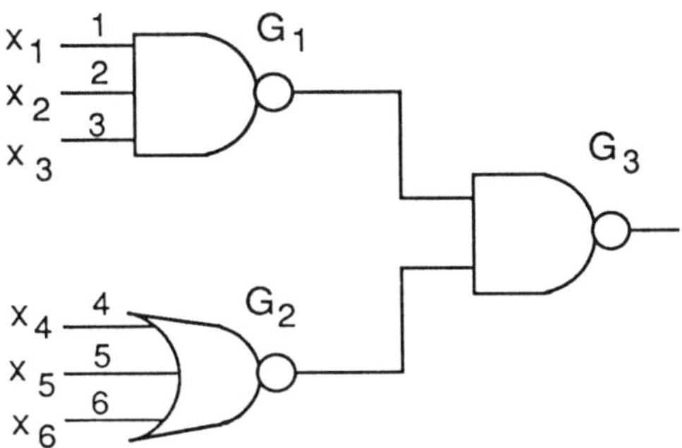

Fig. 4.10 A two-level circuit

can do even better. The two-pattern test $<T_1^1, T_1^0>$ not only detects the stuck-open fault in the nMOS transistor fed by x_1, but also in the other nMOS transistors in series with it which are fed by x_2 and x_3. Thus for checkpoint 2 and 3 we only need two-pattern tests, not three-pattern tests. These two-pattern tests are $<T_2^0, T_2^1>$ and $<T_3^0, T_3^1>$ respectively. However, $T_1^0 = T_2^0 = T_3^0 = 111000$. Thus a test sequence T_{G_1} which detects all the stuck-open faults in gate G_1 is given by $<T_1^0, T_1^1, T_1^0, T_2^1, T_1^0, T_3^1>$. Similarly, we can derive a test sequence T_{G_2} which detects all the stuck-open faults in gate G_2 as $<T_4^1, T_4^0, T_4^1, T_5^0, T_4^1, T_6^0>$. Since T_{G_1} and T_{G_2} detect all the stuck-open faults at the checkpoints, the concatenation of these two sequences gives us the test set for the CMOS circuit. $\square$

For a general CMOS circuit the testing method is as follows. First, a gate-level model is derived for it. This can be done, for example, by modeling at the gate level the nMOS networks of the CMOS gates in the circuit, as explained in the previous section. The primary inputs which do not fan out and the fanout branches in the gate-level model are identified as the checkpoints. The stuck-at $0(1)$ fault test T_i^0 (T_i^1) is derived for checkpoint i. If checkpoint i feeds an AND or NAND gate G_k and checkpoints j_1, j_2, .., j_m also feed G_k then a test sequence T_{G_k} is derived as $<T_i^0, T_i^1, T_i^0, T_{j_1}^1, T_i^0, T_{j_2}^1, .., T_i^0, T_{j_m}^1>$. Similarly, if checkpoint i feeds an OR or NOR gate G_k and checkpoints j_1, j_2, .., j_m also feed G_k then a test sequence T_{G_k} is derived as $<T_i^1, T_i^0, T_i^1, T_{j_1}^0, T_i^1, T_{j_2}^0, .., T_i^1, T_{j_m}^0>$. If the only checkpoint that feeds G_k is checkpoint i then only the first three vectors from T_{G_k} need be fed to it. A NOT gate can be treated as either a one-input NAND gate or NOR gate. After deriving the test sequences for each gate which is fed by at least one checkpoint, the test set is obtained by simply concatenating all these test sequences.

4.1.2 Test Generation at the Switch Level

A test set can also be derived for CMOS circuits from its switch-level or transistor-level description. It is possible to derive tests from the switch-level model for some opens and shorts in MOS circuits which may not be detected if the tests are derived from the gate-level model [GALI80].

Various methods have been presented for CMOS switch-level test generation. We will discuss the methods presented by Chiang and Vranesic in [CHIA83], by Agrawal and Reddy in [AGRA84a], and by Shih and Abraham in [SHIH86].

4.1.2.1 The Chiang-Vranesic Method

Chiang and Vranesic first derive connection graphs of the nMOS and pMOS networks of the CMOS gates and use these graphs for test generation. For example, the connection graphs for the complex gate in Fig. 4.1 is shown in Fig. 4.11. For a pMOS transistor the corresponding arc in the connection graph is labeled with the complement of its input since a pMOS transistor conducts when its input is 0. When a CMOS circuit consists of more than one gate, connection graphs are first derived for each gate and then these graphs are combined into two composite connection graphs for the whole circuit.

A connection graph represents a transmission function. Transmission occurs when there is a conducting path between the source and the sink. For example, let the transmission functions for the connections graphs (a) and (b) in Fig. 4.11 be denoted by f_p and f_n respectively. These functions are given by

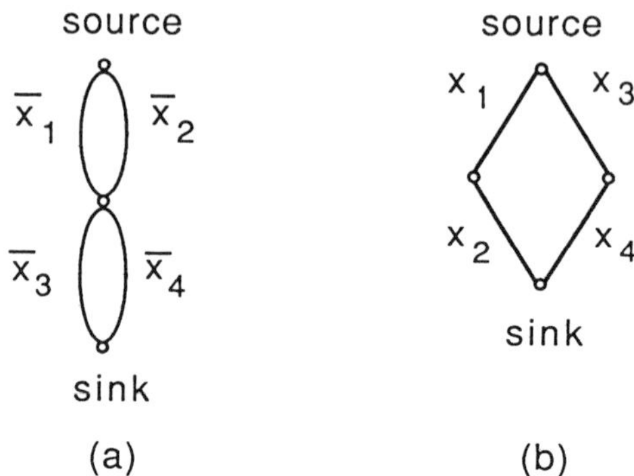

Fig. 4.11 Connection graphs: (a) for pMOS network,
(b) for nMOS network

$$f_p = \bar{x}_1\bar{x}_3 + \bar{x}_1\bar{x}_4 + \bar{x}_2\bar{x}_3 + \bar{x}_2\bar{x}_4$$

$$f_n = x_1x_2 + x_3x_4$$

These transmission functions are always complementary. If f is the function realized by the complex gate then $f = f_p = \bar{f}_n$.

In this method first the test vectors for all stuck-open faults are derived. To detect a stuck-open fault the test vector should activate one or more conduction paths from the source to sink of the relevant connection graph in the fault-free case, but not activate any path when the fault is present. For example, one set of test vectors which can be derived for the connection graph (a) in Fig. 4.11 is $W_1 = \{0101, 1010\}$. For these vectors the function f is 1. Similarly, for the connection graph (b) in Fig. 4.11, we can derive the set $W_0 = \{1100, 0011\}$ for which $f = 0$. In order to provide initialization for the test vectors, the vectors from W_0 and W_1 are selected alternately so that each vector $v_i \in W_0$ is preceded by some vector $v_j \in W_1$ and *vice versa*. For the above example, we can form the sequence $<0101, 1100, 1010, 0011, 0101>$. This forms the test set which detects all the stuck-open faults in the complex gate in Fig. 4.1. When there is more than one gate in the CMOS circuit, one may not be able to guarantee initialization for every stuck-open fault in the circuit by simply deriving an alternating sequence of vectors from W_0 and W_1. Additional two-pattern tests are added to the test sequence for the stuck-open faults for which initialization is not achieved.

Chiang and Vranesic also showed that the test vector for detecting a stuck-open fault in transistor m in a fully complementary CMOS gate is also the test vector for detecting a stuck-on fault in the dual transistor of m. A dual transistor is the transistor in the complementary network which is fed by the same input, or the same branch of the input if the input fans out. For example, a two-pattern test for the stuck-open fault in transistor 1 in the complex gate in Fig. 4.1 is $<0011, 0101>$. The test vector of this two-pattern test is 0101 which also detects a stuck-on fault in transistor 5 which is the dual transistor of transistor 1. Of course, current monitoring may be required to detect the stuck-on fault. This result implies that a test set which detects all stuck-open faults in a CMOS circuit, which has fully complementary gates, also detects all stuck-on faults if current monitoring is done.

4.1.2.2 The Agrawal-Reddy Method

The Agrawal-Reddy method [AGRA84a] is applicable to both nMOS and CMOS circuits and is based on a tree representation of CMOS gates first presented in [LO83]. D-algorithm [ROTH66] is then adapted to the tree representation.

The tree for an MOS gate consists of nodes and branches. There are six basic types of nodes used in [LO83]. However, the Agrawal-Reddy method uses only five types of nodes. These are

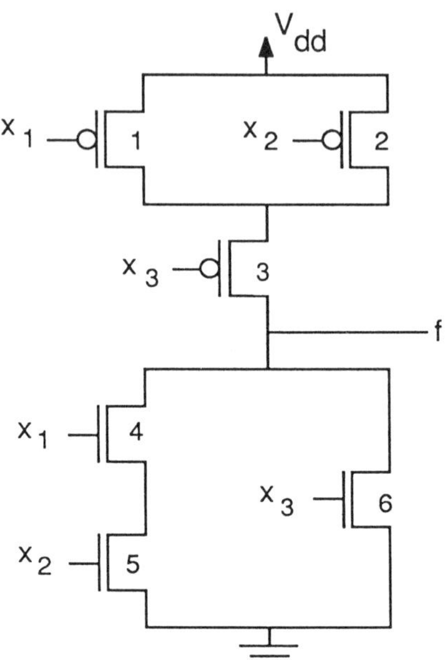

Fig. 4.12 A CMOS complex gate

denoted by T, (*), (+), (=) and P. T is a transistor node, (*) combines transistors in series, (+) combines transistors in parallel and (=) combines the outputs of the nMOS and pMOS sub-trees. P represents a pass transistor. Consider the CMOS gate in Fig. 4.12. Its tree implementation is shown in Fig. 4.13. When there are more than one CMOS gate in the circuit, the equivalent structure for the whole circuit can be derived by simply interconnecting the tree structures obtained for individual CMOS gates.

The tree representation is used for evaluation of the gate output for the given inputs. Note that the input to a T node for a pMOS transistor is complemented. This is required so that both nMOS and pMOS transistors can be uniformly treated. Corresponding to the signals 0, 1, d (don't-care), D and $\overline{\text{D}}$ that are used in D-algorithm, the signals in the tree representation are

OFF, ON, d, SD, $\overline{\text{SD}}$. SD on a branch in the tree implies that the branch is ON (OFF) in the fault-free (faulty) circuit. Similarly, $\overline{\text{SD}}$ on a branch implies that the branch is OFF (ON) in the fault-free (faulty) circuit. Therefore a stuck-open (stuck-on) fault can be represented by SD ($\overline{\text{SD}}$). For an nMOS transistor a D ($\overline{\text{D}}$) at the input results in SD ($\overline{\text{SD}}$) as the branch status, whereas for a pMOS transistor it results in $\overline{\text{SD}}$ (SD) as the branch status. The signal propagation tables for the various nodes are shown in Figs. 4.14-4.18.

When both the nMOS and pMOS networks of a CMOS gate are OFF due to a fault, its output enters into a memory state which is denoted by M in Fig. 4.17. The binary values in the parenthesis in Fig. 4.17 denote the value of M to which the output

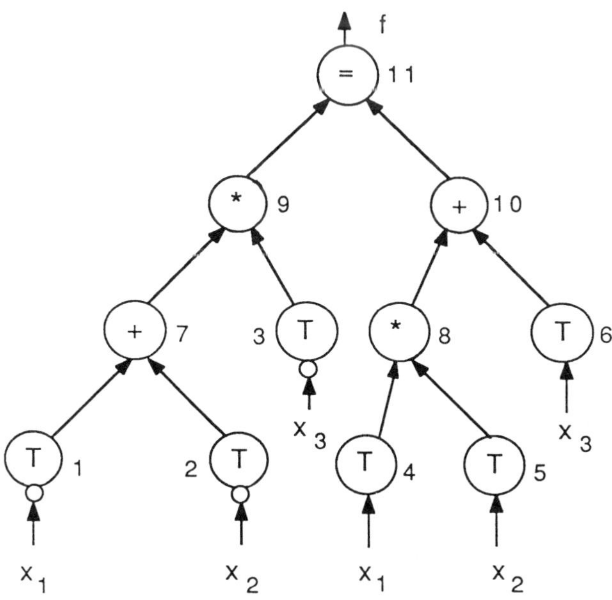

Fig. 4.13 The tree representation

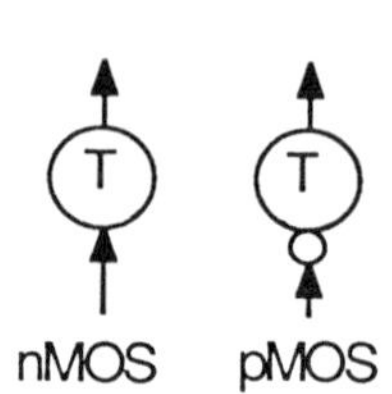

Input	Output
0	OFF
1	ON
D	SD
$\overline{D}$	$\overline{SD}$
d	d

Fig. 4.14 T node and its evaluation table

*	OFF	ON	SD	$\overline{SD}$	d
OFF	OFF	OFF	OFF	OFF	OFF
ON	OFF	ON	SD	$\overline{SD}$	d
SD	OFF	SD	SD	OFF	d
$\overline{SD}$	OFF	$\overline{SD}$	OFF	$\overline{SD}$	d
d	OFF	d	d	d	d

Fig. 4.15 (*) node and its evaluation table

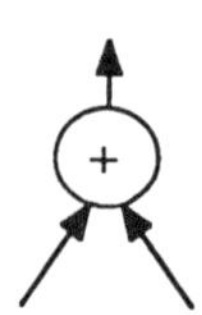

+	OFF	ON	SD	$\overline{SD}$	d
OFF	OFF	ON	SD	$\overline{SD}$	d
ON	ON	ON	ON	ON	ON
SD	SD	ON	SD	ON	d
$\overline{SD}$	$\overline{SD}$	ON	ON	$\overline{SD}$	d
d	d	ON	d	d	d

Fig. 4.16 (+) node and its evaluation table

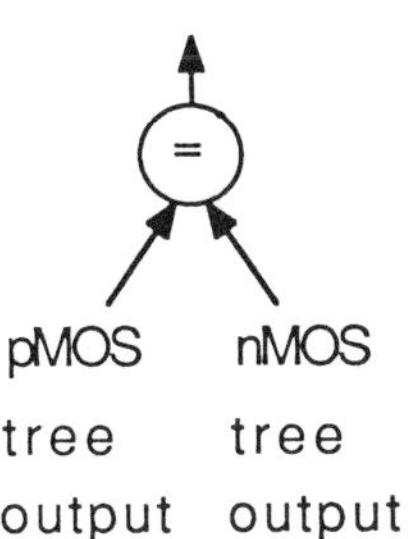

=	OFF	ON	SD	$\overline{SD}$	d
OFF	M	0	$\overline{D}(1)$	$D(1)$	d
ON	1	0	$\overline{D}$	D	d
SD	$D(0)$	0	$\overline{D}(1)$	D	d
$\overline{SD}$	$\overline{D}(0)$	0	$\overline{D}$	$D(1)$	d
d	d	0	d	d	d

Fig. 4.17 (=) node and its evaluation table, nMOS (pMOS) tree output given horizontally (vertically)

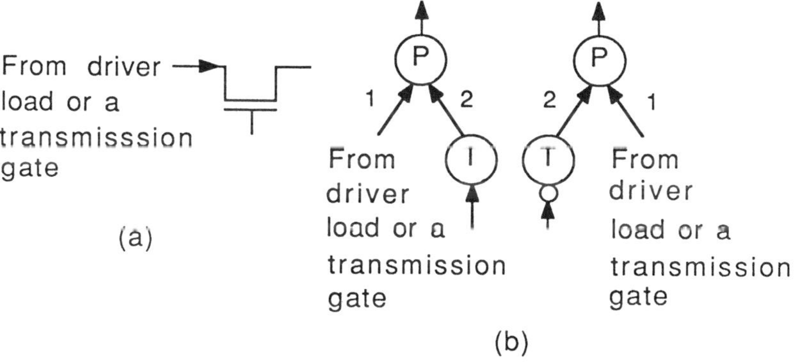

(b)

Input 2 →	Previous output = 0			Previous output = 1		
Input 1 ↓	OFF	ON	d	OFF	ON	d
0	0^*	0	0^*	1^*	0	d
1	0^*	1	d	1^*	1	1^*
0^*	0^*	0^*	0^*	1^*	d	d
1^*	0^*	d	d	1^*	1^*	1^*
d	0^*	d	d	1^*	d	d

(c)

Fig. 4.18 (a) A transmission gate, (b) P node representation of an nMOS and pMOS transistor, (c) evaluation table

of the CMOS gate has to be initialized in order to detect the corresponding fault. For example, if the output of the pMOS (nMOS) sub-tree is OFF (SD) then the output of the (=) node and hence the output of the CMOS gate is $\overline{D}(1)$. This means that a $\overline{D}$ is generated at the output only if it was initialized to 1. In the table in Fig. 4.17 it has also been assumed that if both the networks of a CMOS gate conduct then the nMOS network dominates and the output is 0. This assumption considerably simplifies the model.

In case of a transmission gate, the source and drain can interact when the input is ON. To comprehensively model this effect the previous values on both the source and drain should be taken into account. However, to avoid complexity, the somewhat simplified evaluation table in Fig. 4.18 is used. In this table 0^* and 1^* represent 0 and 1 logic values at high impedance nodes. Also to be kept in mind is that this evaluation table is used once for the fault-free circuit and then for the faulty circuit to determine the combined values (i.e. to find if the output is D or $\overline{D}$).

This method is aimed at test generation for the stuck-at, stuck-open and stuck-on faults. In D-algorithm a D or $\overline{D}$ is injected at the fault site and the effect is propagated to a circuit output through D-drive (see Section 3.1). The line justification is done to justify the logic values on the different lines which are specified by the D-drive. Backtracking may be necessary before a test vector can be generated. This approach can be adapted to the tree representation as well. This can be illustrated through an example.

Example 4.6: Suppose that we want to derive a two-pattern test for the stuck-open fault in transistor 4 in the complex gate in Fig. 4.12. We will first derive the test vector. The input x_1 must be 1 which results in the error signal SD at the output of the T node corresponding to transistor 4, as shown in Fig. 4.19. To propagate the effect through node 8 in Fig. 4.19, the output of node 5 should

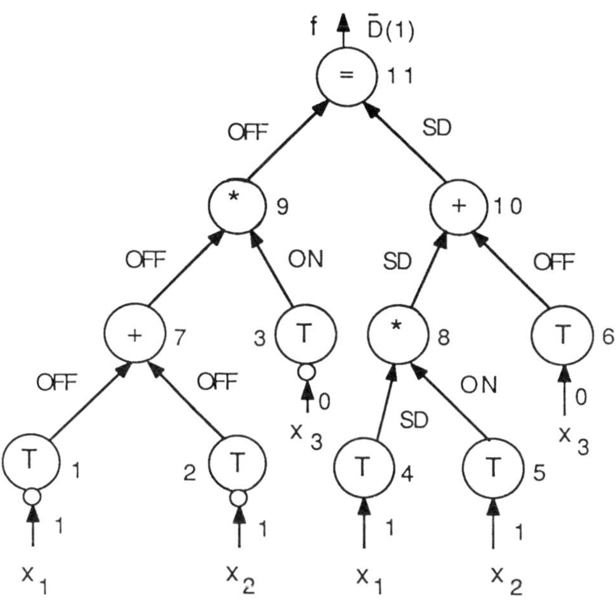

Fig. 4.19 Test generation example

be ON from the evaluation table in Fig. 4.15. To further propagate the effect through node 10, the output of node 6 must be
OFF from the evaluation table in Fig. 4.16. The above conditions
imply that $x_2 = 1$ and $x_3 = 0$. Thus the vector that has been
obtained is 110. This results in an OFF signal at the output of
node 9 as shown in Fig. 4.19. The final output f is then found to
be $\overline{D}(1)$ from the evaluation table in Fig. 4.17. This means that
the output is $\overline{D}$ if it is first initialized to 1. One initialization vector which results in $f = 1$ is 010. Thus $<010,110>$ is a two-

pattern test for the stuck-open fault. □

4.1.2.3 The Shih-Abraham Method

Shih and Abraham [SHIH86] also proposed a method similar to the Agrawal-Reddy method based on the tree structure proposed in [LO83]. They recognized that since the transmission function of the pMOS network of a CMOS gate is always complementary to the transmission function of its nMOS network (see Section 4.1.2.1), it is wasteful of computation to evaluate both these functions separately. They introduced another node called the negation node N for rapid error propagation through fault-free CMOS gates. For example, if the complex gate in Fig. 4.12 were a fault-free CMOS gate embedded in a CMOS circuit through which only error propagation was desired, then its tree representation could be obtained as shown in Fig. 4.20. The evaluation table for node N is given in Fig. 4.21. We have employed the notation used in Section 4.1.2.2. Only if a fault is to be injected in the CMOS gate is there a need to use the full tree representation shown in Fig. 4.13.

Shih and Abraham also used this technique to generate tests for some bridging faults.

4.2 ROBUST TEST GENERATION

Robust test generation for static CMOS circuits has been considered in [REDD84b, AGRA84b, REDD85, WEIW86, JHA88, JHA89]. In robust testing it is assumed that the primary inputs of the circuit do not have any static hazards or glitches. This is a reasonable assumption in most cases. However, timing skews among two or more inputs are allowed. Furthermore, arbitrary

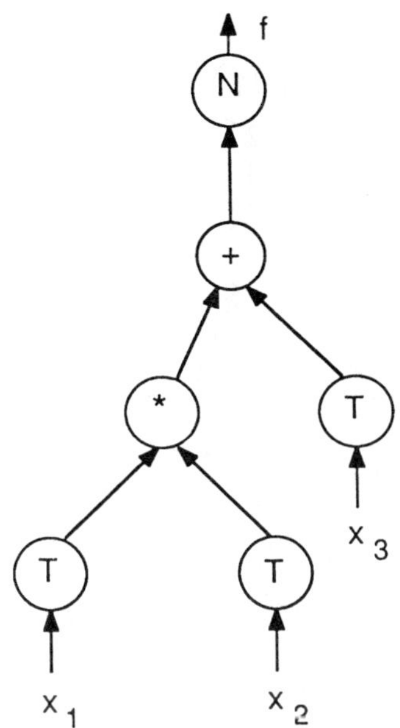

Fig. 4.20 Tree representation for a fault-free gate

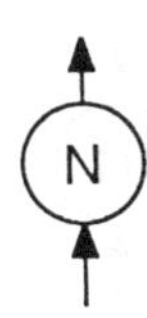

Input	Output
OFF	1
ON	0
SD	$\overline{D}$
$\overline{SD}$	D
d	d

Fig. 4.21 N node and its evaluation table

delays are assumed to be present within the circuit. We will present in this section the method given by Reddy, Reddy and

Agrawal in [REDD84b].

4.2.1 The Reddy-Reddy-Agrawal Method

Reddy *et al.* use the gate-level model of Jain and Agrawal [JAIN83, JAIN85], which was introduced in Section 4.1.1.1, for robust test generation for static CMOS circuits consisting of an interconnection of fully complementary gates. By this method a robust two-pattern test can be derived for a stuck-open fault if such a test exists. The approach taken in this method is to first generate the test vector T_2 for the stuck-open fault and then try to generate an initialization vector T_1 such that the two-pattern test $<T_1,T_2>$ is robust.

Test invalidation occurs when a static hazard is present at the output of the faulty CMOS gate. A static 0-hazard (1-hazard) is said to be present on line k for a transition from T_1 to T_2 if and only if (a) the expected logic value on line k under T_1 and T_2 is 0(1), and (b) during the transition from T_1 to T_2 it is possible that line k can momentarily take the logic value 1(0) due to circuit delays and/or timing skews at the primary inputs.

Let T_1 be applied at time t_1 and T_2 at time t_2. From Section 4.1.1.1 we may recall that the Jain-Agrawal logic model of a CMOS gate consists of gate-level models of its pMOS and nMOS networks which are connected to a B-block. The inputs to the B-block are S_1 and S_0. When a stuck-open fault is present, both S_1 and S_0 are 0 at time t_2 indicating that there is no conduction path through either the pMOS or the nMOS network. Reddy *et al.*

proved the following theorem:

Theorem 4.1: A two-pattern test $<T_1,T_2>$ is robust for detecting a stuck-open fault in an nMOS (pMOS) transistor in gate G of a CMOS circuit if and only if the S_0 (S_1) input of the B-block modeling the output of G is free of static 0-hazard.

This theorem tells us that given T_2, T_1 should be derived such that $<T_1,T_2>$ does not result in a static 0-hazard on S_0 or S_1 as the case may be. We associate a sequence of two values from $\{0,1,d\}$ with every line which is fed these values at t_1 and t_2. We also associate a hazard status with a line which can be from the set $\{hp,hu,hf\}$, where hp denotes hazard present, hu denotes hazard unknown and hf denotes hazard-free. LHF denotes the set of lines which are required to be hazard-free at some step in the algorithm.

We next need to determine how we can specify the inputs of AND, OR, NOT gates and the B-block to satisfy a hazard-free requirement on their outputs. This is shown in Table 4.5.

We also need to determine how the hazard status of the output of a gate can be found given the hazard status of its inputs. Consider the gates and their inputs and output as shown in Fig. 4.22. x (y) for the NOT gate and v, x_i (w, y_i), $i = 1,2,..,n$, for AND and OR gates denote the logic values on the lines at time t_1 (t_2), whereas h, h_1, h_2, .., h_n denote their hazard status. We know that x, y, x_i, y_i, v, w $\in \{0,1,d\}$ and h, $h_i \in \{hp,hu,hf\}$. For the NOT gate the hazard status of the output is the same as that of its input. For an AND gate, h can be computed as follows:

(1)　If $v = w = 0$ then h = hf, if and only if at least one input is
00hf

Table 4.5 Satisfying hazard-free requirements

Gate	Requirement at the output	Status of the inputs
NOT	0-hazard-free	1-hazard-free
NOT	1-hazard-free	0-hazard-free
OR	0-hazard-free	all inputs 0-hazard-free
OR	1-hazard-free	at least one input 1-hazard-free
AND	0-hazard-free	at least one input 0-hazard-free
AND	1-hazard-free	all inputs 1-hazard-free
B-block	0-hazard-free	S_1 0-hazard-free, S_0 1 at t_1
B-block	1-hazard-free	S_0 0-hazard-free, S_1 1 at t_1

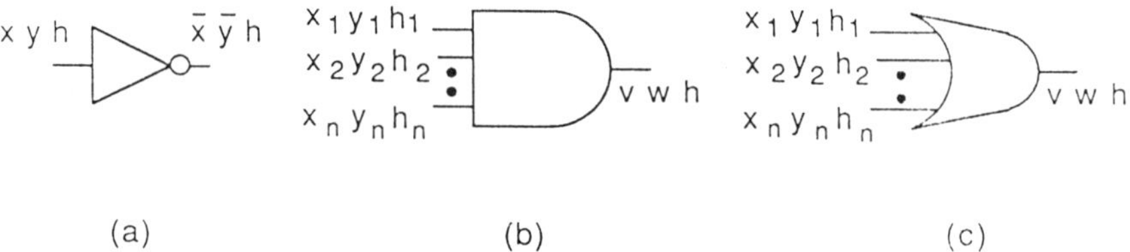

Fig. 4.22 Different primitive gates

$= hu$, if and only if no input is 00hf and at least one is 00hu, 0dhu, d0hu or ddhu

$= hp$, otherwise

(2) If $v = w = 1$ then $h = hf$, if and only if $h_1 = h_2 = .. = h_n = hf$

$= hp$, if and only if at least one of $\{h_1, h_2,..,h_n\}$ is hp

$$= hu, \text{ otherwise}$$

(3) If $v = w = d$ or $v \neq w$ then $h = hu$.

Similarly, for an OR gate, h can be computed as follows:

(1) If $v = w = 1$ then $h = hf$, if and only if at least one input is 11hf

$= hu$, if and only if no input is 11hf and at least one is 11hu, 1dhu, d1hu or ddhu

$= hp$, otherwise

(2) If $v = w = 0$ then $h = hf$, if and only if $h_1 = h_2 = .. = h_n = hf$

$- hp$, if and only if at least one of $\{h_1, h_2,..,h_n\}$ is hp

$= hu$, otherwise

(3) If $v = w = d$ or $v \neq w$ then $h = hu$.

Finally, hazard status computation for the B-block is shown in Table 4.6. Note that the hazard status of S_1 and S_0 are always identical for fully complementary CMOS gates.

We are now in a position to see how T_1 can be derived, given T_2, such that $<T_1,T_2>$ is robust. The procedure for deriving T_1 is quite similar to the line justification step in D-algorithm. Let G be the logic model derived by the Jain-Agrawal method for the CMOS circuit. For deriving T_1 only that part of G is used which includes the lines that drive the gate under test. Let this part be called G'. We know from Section 4.1.1.1 that a stuck-open fault in the nMOS (pMOS) transistor in a CMOS gate can be translated

Table 4.6 Hazard status computation for the B-block

S_1	S_0	Output		S_1	S_0	Output
11hf	00hf	11hf		00hp	11hp	00hp
11hp	00hp	11hp		00hu	11hu	00hu
11hu	00hu	11hu		0dhu	1dhu	0dhu
10hu	01hu	10hu		d1hu	d0hu	d1hu
1dhu	0dhu	1dhu		d0hu	d1hu	d0hu
01hu	10hu	01hu		ddhu	ddhu	ddhu
00hf	11hf	00hf				

into a stuck-at 0(1) fault in G'. A hazard-free requirement is assumed to be automatically satisfied upon its arrival at a primary input. Thus a primary input need not be assigned the hazard status of hf *a priori*.

Initially the set LHF is set to null. The state of the faulty line is set to its stuck value and hazard status hf. The state of all the other lines is set to d at time t_1 and hazard status hu. The S_1 (S_0) line of the B-block of the faulty gate is added to LHF if the fault under test is in a pMOS (nMOS) transistor. S_1 (S_0) is set to 00hu (10hu) for a pMOS transistor stuck-open fault and 10hu (00hu) for an nMOS transistor stuck-open fault. If LHF is non-empty, an element m of LHF is picked. Then an attempt is made using Table 4.5 to specify the inputs of the gate or B-block whose output is m such that the hazard status of m becomes hf. If successful, m is deleted from LHF. Otherwise, backtracking is done to another choice, if possible, and LHF is updated. Line justification is done to satisfy the new requirements. In the execution of the above steps some gate inputs or outputs may be specified such that they uniquely imply values on other lines in the circuit. Implication can be made both in the forward and backward direction. If in the process of implication the hazard status of a line in LHF becomes hf then the line is deleted from LHF. If a conflict occurs in logic values or hazard status during implication, then backtracking is done to the last choice made, if possible. The vector T_1 is obtained when LHF becomes empty and no line justifications are to be done. Otherwise, the above process is repeated.

The working of this method should be clearer through an example.

Example 4.7: Consider the CMOS complex gate in Fig. 4.12 whose logic model is shown in Fig. 4.23. Suppose we want to robustly detect the stuck-open fault in transistor 1 in Fig. 4.12.

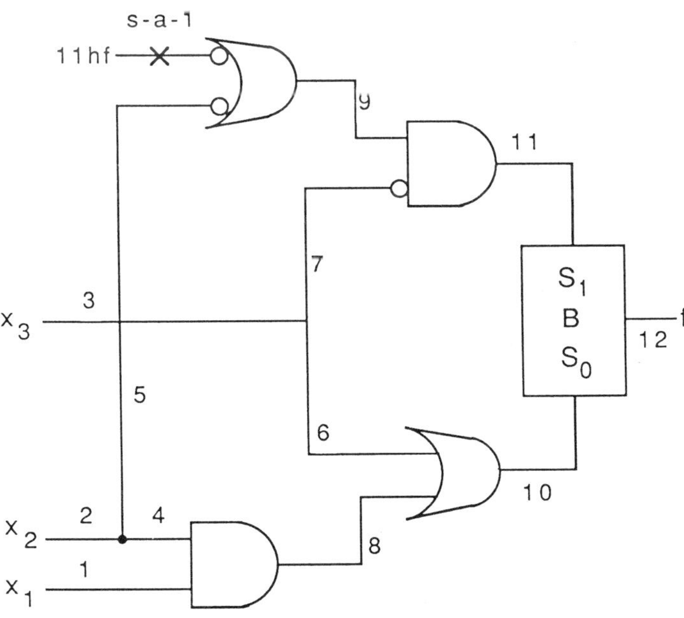

Fig. 4.23 The logic model

The corresponding faulty line in Fig. 4.23 is isolated and assigned 11hf. A test vector T_2 can be derived by any of the previous methods to be 010. The different steps in the procedure for deriving T_1, given T_2, is shown in Table 4.7. In this table a '-' entry indicates that the previous assignment to the line remains valid. From this table, T_1 can be derived as d11. □

4.2.2 Some Issues in Robust Test Generation

The disadvantage of using a robust test generation algorithm is that it can substantially increase the computer time required for test generation. The reason for this, of course, is the care that has

Table 4.7 Derivation of T_1

Line number	Initial value	Step 1	Step 2	Step 3	Step 4
1	d0hu	-	-	-	-
2	d1hu	-	-	-	11hf
3	d0hu	-	-	10hu	-
4	d1hu	-	-	-	-
5	d1hu	-	-	11hu	11hf
6	d0hu	-	10hu	-	-
7	d0hu	-	-	10hu	-
8	d0hu	-	-	-	-
9	d0hu	-	00hu	00hf	-
10	d0hu	10hu	-	-	-
11	d0hu	00hu	00hf	-	-
12	d0hu	00hf	-	-	-
LHF	null	{11}	{9}	{5}	

to be taken in deriving the initialization vector T_1 given the test vector T_2. Another problem that arises is that any test set derived from robust two-pattern tests for the stuck-open faults at the checkpoints in an irredundant static CMOS circuit is not guaranteed to robustly detect all the stuck-open faults in the circuit. Recall from Section 4.1.1.3 that the checkpoints of a CMOS circuit are its primary inputs which do not fan out and all the fanout branches. Consider the irredundant NAND-NAND CMOS circuit whose gate-level model is shown in Fig. 4.24. In this circuit the checkpoints are the four primary inputs. There are 8 transistors associated with these checkpoints. A test sequence which contains a robust two-pattern test for stuck-open faults in each of these 8 transistors is <1101, 0101, 1110, 1010, 0111, 0101, 1011, 1010>. This sequence does not robustly detect the stuck-open faults in either the pMOS transistor fed by f_1 or the pMOS transistor fed by f_2. However, other test sequences can be derived from robust two-pattern tests obtained for stuck-open faults at the checkpoints which also robustly detect all the stuck-open faults in the circuit. One such test sequence is <1101, 0101, 1101, 1001, 0111, 0101, 0111, 0110>. This example tells us that for robust testing we also have to concentrate on stuck-open faults

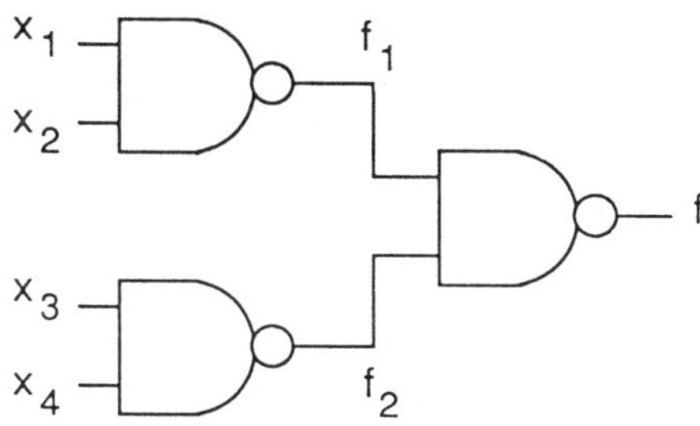

Fig. 4.24 A NAND-NAND circuit

which are not necessarily located at the checkpoints. This greatly increases the number of stuck-open faults that have to be considered and consequently increases the test generation time.

In order to increase the possibility of robustly detecting stuck-open faults, which are not located at the checkpoints, heuristics may be used. For example, by reducing the Hamming distance of the two vectors (number of bits they differ in) in a two-pattern test we reduce the possibility of test invalidation. The best we can do is to reduce the Hamming distance to 1. In the above example, the second sequence consists of the following six two-pattern tests: $<1101, 0101>$, $<0101, 1101>$, $<1101, 1001>$, $<0111, 0101>$, $<0101, 0111>$ and $<0111, 0110>$. In each two-pattern test the two vectors differ in only one bit. In the first test sequence this condition was not met by every two-pattern test. Therefore the second test sequence turned out to be superior. However, it has not yet been proved that if robust two-pattern tests are derived from the checkpoints of an irredundant CMOS circuit such that the two vectors in the tests differ in only one bit, then the test set derived from these two-pattern tests also robustly detects all the stuck-open faults in the circuit. Some irredundant CMOS circuits exist for which it is not possible to derive robust two-pattern tests, in which the two vectors differ in only one bit, for every checkpoint in the circuit. However, such circuits are relatively rare. A method for deriving two-pattern tests from the checkpoints in which the two vectors have Hammming distance 1, when such tests exist, is given in [JHA88, JHA89]. If we do not want to worry about this problem, we can use the robustly testable CMOS circuit designs presented in the next chapter.

REFERENCES

[AGRA84a] P. Agrawal and S. M. Reddy, "Test generation at MOS level," in *Proc. Int. Conf. Computers, Systems & Signal Processing*, Bangalore, India, Dec. 1984.

[AGRA84b] P. Agrawal, "Test generation at the switch level," in *Proc. Int. Conf. Computer-Aided Design*, Santa Clara, CA, pp. 128-130, Nov. 1984.

[CHAN83] R. Chandramouli, "On testing stuck-open faults," in *Proc. Int. Symp. Fault-Tolerant Comput.*, Milan, Italy, pp. 258-265, June 1983.

[CHIA83] K. Chiang and Z. G. Vranesic, "On fault detection in CMOS logic networks," in *Proc. Design Automation Conf.*, Miami Beach, FL, pp. 50-55, June 1983.

[GALI80] J. Galiay, Y. Crouzet, and M. Vergniault, "Physical versus logical fault models in MOS LSI circuits: Impact on their testability," *IEEE Trans. Comput.*, vol. C-29, pp. 527-531, June 1980.

[JAIN83] S. K. Jain and V. D. Agrawal, "Test generation for MOS circuits using D-Algorithm," in *Proc. Design Automation Conf.*, Miami Beach, FL, pp. 64-70, June 1983.

[JAIN85] S. K. Jain and V. D. Agrawal, "Modeling and test generation algorithms for MOS circuits," *IEEE Trans. Comput.*, vol. C-34, pp. 426-433, May 1985.

[JHA88] N. K. Jha, "Multiple stuck-open fault detection in CMOS logic circuits," *IEEE Trans. Comput.*, vol. 37, pp. 426-432, Apr. 1988.

[JHA89] N. K. Jha, "Robust testing of CMOS logic circuits," *Int. J. Computers & Electrical Engg.*, vol. 15, no. 1, pp. 19-28,

1989.

[KOHA78] Z. Kohavi, Switching and Finite Automata Theory, McGraw-Hill, 1978.

[LO83] C. Y. Lo, H. N. Nham, and A. K. Bose, "A data structure for MOS circuits," in *Proc. Design Automation Conf.*, Miami Beach, FL, pp. 619-624, June 1983.

[REDD83] S. M. Reddy, M. K. Reddy, and J. G. Kuhl, "On testable design for CMOS logic circuits," in *Proc. Int. Test Conf.*, Philadelphia, PA, pp. 435-445, Oct. 1983.

[REDD84a] S. M. Reddy, V. D. Agrawal, and S. K. Jain, "A gate-level model for CMOS combinational logic circuits with application to fault detection," in *Proc. Design Automation Conf.*, Albuquerque, NM, pp. 504-509, June 1984.

[REDD84b] S. M. Reddy, M. K. Reddy, and V. D. Agrawal, "Robust tests for stuck-open faults in CMOS combinational logic circuits," in *Proc. Int. Symp. Fault-Tolerant Comput.*, Orlando, FL, pp. 44-49, June 1984.

[REDD85] M. K. Reddy, S. M. Reddy, and P. Agrawal, "Transistor level test generation for MOS circuits," in *Proc. Design Automation Conf.*, Las Vegas, pp. 825-828, June 1985.

[REDD86] S. M. Reddy and M. K. Reddy, "Testable realizations for FET stuck-open faults in CMOS combinational logic circuits," *IEEE Trans. Comput.*, vol. C-35, pp. 742-754, Aug. 1986.

[ROTH66] J. P. Roth, "Diagnosis of automata failures: A calculus and a method," *IBM J. Res. & Dev.*, vol. 10, no. 4, pp. 278-291, July 1966.

[SHIH86] H. Shih and J. A. Abraham, "Transistor level test generation for physical failures in CMOS circuits," in *Proc. Design Automation Conf.*, Las Vegas, pp. 243-249, June 1986.

[WEIW86] M. Weiwei and L. Xieting, "Robust test generation algorithm for stuck-open fault in CMOS circuits," in *Proc.*

Design Automation Conf., Las Vegas, pp. 236-242, June 1986.

ADDITIONAL READING

[AL-AR87] S. A. Al-Arian and D. P. Agrawal, "Physical failures and fault models of CMOS circuits," *IEEE Trans. Circuits & Systems*, vol. CAS-34, no. 3, pp. 269-279, Mar. 1987.

[CHEN84] H. H. Chen, R. G. Mathews, and J. A. Newkirk, "Test generation for MOS circuits," in *Proc. Int. Test Conf.*, Philadelphia, PA, pp. 70-79, Oct. 1984.

[COX88a] H. Cox and J. Rajski, "A method of fault analysis for test generation and fault diagnosis," *IEEE Trans. CAD*, vol. 7, pp. 813-833, July 1988.

[COX88b] H. Cox and J. Rajski, "Stuck-open and transition fault testing in CMOS complex gates," in *Proc. Int. Test Conf.*, Washington, D.C., pp. 688-694, Sept. 1988.

[El-ZI81a] Y. M. El-Ziq, "Automatic test generation for stuck-open faults in CMOS VLSI," in *Proc. Design Automation Conf.*, Nashville, TN, pp. 347-354, June 1981.

[El-ZI81b] Y. M. El-Ziq and R. J. Cloutier, "Functional-level test generation for stuck-open faults in CMOS VLSI," in *Proc. Int. Test Conf.*, Philadelphia, PA, pp. 536-546, Oct. 1981.

[GUPT88] G. Gupta and N. K. Jha, "A universal test set for CMOS circuits," *IEEE Trans. CAD*, vol. 7, pp. 590-597, May 1988.

[JHA86] N. K. Jha, "Detecting multiple faults in CMOS circuits," in *Proc. Int. Test Conf.*, Washington, D.C., pp. 514-519, Sept. 1986.

[KOE88] W.-Y. Koe and S. F. Midkiff, "Circuit simulation of CMOS faults," in *Proc. SOUTHEASTCON*, Knoxville, TN, pp. 87-91, Apr. 1988.

[MALA82] Y. K. Malaiya and S. Y. H. Su, "A new model and testing technique for CMOS devices," in *Proc. Int. Test Conf.*, Philadelphia, PA, pp. 25-34, Oct. 1982.

[MALY88] W. Maly and P. Nigh, "Built-in current testing - feasibility study," in *Proc. Int. Conf. Computer-Aided Design*, Santa Clara, CA, pp. 340-343, Nov. 1988.

[MAND84] K. D. Mandl, "CMOS VLSI challenges to test," in *Proc. Int. Test Conf.*, Philadelphia, PA, pp. 642-648, Oct. 1984.

[RAJS86] J. Rajski and H. Cox, "Stuck-open fault testing in large CMOS networks by dynamic path tracing," in *Proc. IEEE Int. Conf. Computer Design*, Port Chester, NY, pp. 252-255, Oct. 1986.

[RAJS89] R. Rajsuman, Y. K. Malaiya, and A. P. Jayasumana, "Limitations of switch-level analysis for bridging faults," *IEEE Trans. CAD*, vol. 8, pp. 807-811, July 1989.

[ROBI85] S. Robinson and J. Shen, "Towards a switch-level test pattern generation program," in *Proc. Int. Conf. Computer-Aided Design*, Santa Clara, CA, pp. 39-41, Nov. 1985.

[SHIH86] H.-C. Shih and J. A. Abraham, "Fault collapsing techniques for MOS VLSI circuits," in *Proc. Int. Symp. Fault-Tolerant Comput.*, Vienna, Austria, pp. 370-375, June 1986.

PROBLEMS

4.1. The gate-level model for the complex gate in Fig. 4.4 can be derived from the circuits shown in Figs. 4.8 and 4.9 using the Reddy-Agrawal-Jain method. Use this gate-level model to derive a two-pattern test to detect a stuck-open fault in transistor 4 in the complex gate.

4.2. Prove Result 4.1.

4.3. Use the Chandramouli method to derive a test set for detecting all the stuck-open faults in the NAND-NAND CMOS circuit which implements the function $f = x_1x_2 + \overline{x}_2x_3 + \overline{x}_1\overline{x}_3$.

4.4. Prove that a stuck-open fault test set derived for an irredundant static CMOS circuit, which is formed by interconnecting fully complementary gates, also detects all the stuck-on faults in it using current monitoring.

4.5. Using the Agrawal-Reddy method, obtain a tree representation for the CMOS circuit shown in Fig. 4.25.

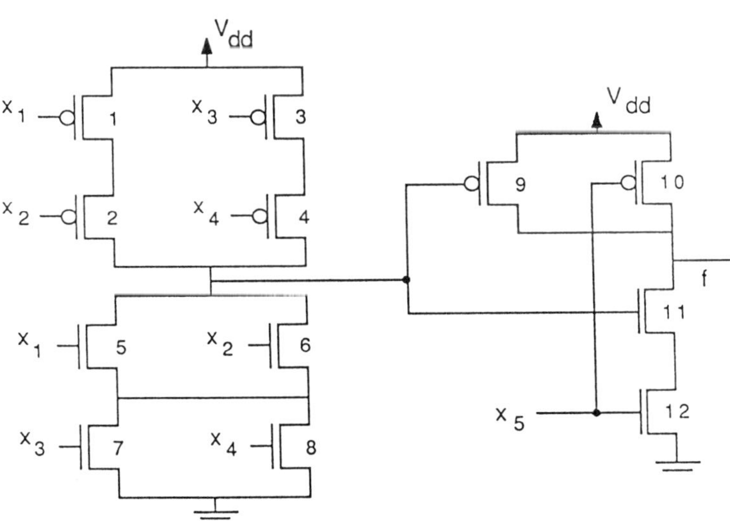

Fig. 4.25 A static CMOS circuit

From this tree representation:

(a) Derive a two-pattern test for the stuck-open fault in transistor 2.

(b) Derive a test to detect the stuck-on fault in transistor 9. Assume that current monitoring is done to detect the fault.

4.6. Obtain the tree representation for detecting the stuck-open fault in transistor 6 in the circuit in Fig. 4.25 using the Shih-Abraham method. Derive a two-pattern test for this fault.

4.7. Prove Theorem 4.1.

4.8. Derive the logic model for the circuit in Fig. 4.25 using the Jain-Agrawal method. Use this model to derive a robust two-pattern test for the stuck-open fault in transistor 7.

4.9. Obtain an irredundant static CMOS circuit in which it is not possible to derive a two-pattern test for every stuck-open fault at the checkpoints, such that the two vectors of the two-pattern test differ in only one bit position.

4.10. Develop a method for deriving a two-pattern test for a stuck-open fault at a checkpoint in any static CMOS circuit such that the two vectors differ in only one bit position, whenever it is possible to do so.

Chapter 5
DESIGN FOR ROBUST TESTABILITY

If a test set exists for a CMOS circuit which detects all the stuck-open faults in it in the presence of arbitrary circuit delays and timing skews, then the circuit is said to be robustly testable. We saw earlier in Chapter 2 that dynamic CMOS circuits, such as domino CMOS and cascode voltage switch logic circuits, are always robustly testable. However, this is not true for all static CMOS circuits. In this chapter, procedures for designing robustly testable static CMOS circuits will be discussed. Generally, the circuits are designed to be robustly testable with respect to single stuck-open faults. However, we will also discuss a method which guarantees robust testability with respect to multiple faults.

We know that stuck-open faults in static CMOS circuits require two-pattern tests. Similar tests are also required to detect delay faults. We will see that the test sets derived for these two fault models are related in some cases.

Many methods have been presented in literature for designing robustly testable static CMOS circuits [REDD83, JHA84, JHA85, REDD86, LIU87, CRAI87, SHER88, KUND88a, KUND88b, KUND89]. Each technique has its advantages and disadvantages. In choosing one technique over the other, a designer is faced with certain trade-offs. Some of the issues involved are:

(1)   Is it permissible to use extra controllable inputs?

(2)   Can one use extra observation points (additional outputs)?

(3)   Is there any restriction on the type of gates that may be used?

(4)   What are the fan-in and fan-out allowed?

(5)   What is the permissible logic overhead for obtaining a robustly testable design?

We begin by discussing design techniques which require additional inputs.

5.1 TESTABLE DESIGNS USING EXTRA INPUTS

This approach is taken in [REDD83, REDD86, LIU87]. The method given in [REDD83, REDD86] is primarily applicable to two-level circuits, although robustly testable multi-level circuits can also be obtained from the two-level circuits to meet the fan-in constraints. It is a well-known fact that any arbitrary function f can be realized by a two-level NAND-NAND or NOR-NOR circuit [KOHA78]. If NAND gates are used then the function f is realized as a sum-of-products expression. Otherwise, if NOR gates are used then the function f is realized as a product-of-sums expression.

As mentioned in Section 4.2 in the last chapter, for robustly testable designs it is usually assumed that the primary inputs of the circuit do not have any static hazards or glitches. In most cases this is a reasonable assumption. However, no assumption is made regarding the time of arrival of any two inputs with respect to each other. In other words, timing skews among two or more inputs is allowed. Furthermore, arbitrary delays within the circuit are assumed to be present.

We will first discuss the method given in [REDD83, REDD86].

5.1.1 The Reddy-Reddy-Kuhl Method

Let us first consider an example of a two-level circuit which is not robustly testable with respect to all its single stuck-open faults. Refer to the NAND-NAND circuit shown in Fig. 5.1. This can be viewed as the gate-level model of the NAND-NAND CMOS circuit which implements the function $f = x_3\overline{x}_4x_5 + x_1x_2x_3 + x_3x_4\overline{x}_5 + x_1x_3\overline{x}_5 + \overline{x}_1\overline{x}_2\overline{x}_4 + \overline{x}_3\overline{x}_4\overline{x}_5$. Consider a stuck-open fault in the pMOS transistor of gate G_7 in the circuit which is driven by the output of gate G_4. In order to detect this fault we first need to

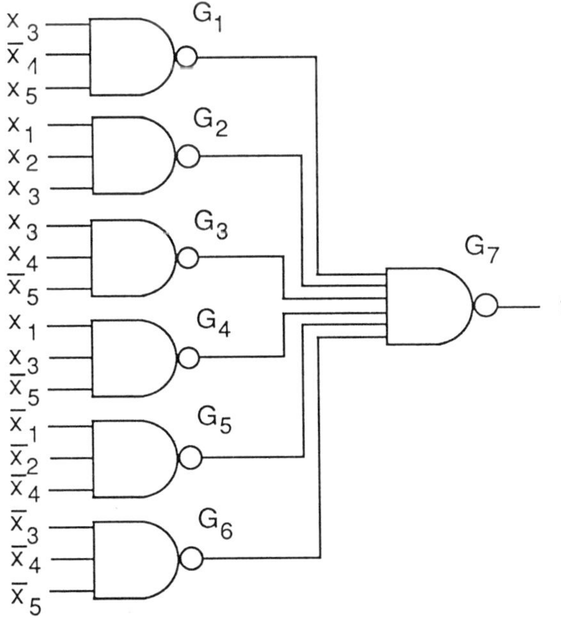

Fig. 5.1 A NAND-NAND circuit

initialize the output f to 0. There are fifteen vectors which can accomplish this. They are 00010, 00011, 00111, 01001, 01010, 01011, 01100, 01111, 10001, 10010, 10011, 10111, 11001, 11010 and 11011. We then need to apply a test vector which tries to activate a conduction path through the faulty transistor. The only test vector possible is 10100. Thus there are fifteen different two-pattern tests possible. However, for each one of these tests a conduction path in parallel to the one containing the faulty transistor may get activated in the presence of circuit delays and timing skews. Thus none of the possible two-pattern tests is robust.

The design technique in this approach is to first determine if the two-level CMOS circuit realization of the function is robustly testable. If not, then the circuit is made robustly testable by adding an extra input. In order to determine if a two-level circuit is robustly testable, the following results are very useful.

Result 5.1: For an irredundant two-level NAND-NAND CMOS circuit a robust two-pattern test exists for all single stuck-open faults except possibly for those in the pMOS network of the output NAND gate.

Result 5.2: For an irredundant two-level NOR-NOR CMOS circuit a robust two-pattern test exists for all single stuck-open faults except possibly for those in the nMOS network of the output NOR gate.

Results 5.1 and 5.2 reduce the search effort by limiting the number of transistors that have to be examined. The two results are also dual of each other. Thus, henceforth, we will concentrate on only NAND-NAND CMOS circuits. The results that are obtained for these circuits can be trivially extended to NOR-NOR CMOS circuits. The above results were later generalized to multi-output circuits in [CRAI87].

A vector is sometimes also referred to as a vertex. A vertex for which a function assumes the value 0(1) is called a 0(1)-vertex. Two vertices or vectors are said to be adjacent if they differ in only one bit position. For example, 1010 and 1011 are adjacent, but 1010 and 1001 are not. The following result pertains to the pMOS transistors in the output NAND gate of a NAND-NAND CMOS circuit.

Result 5.3: A robust two-pattern test for a stuck-open fault in a pMOS transistor of the output NAND gate of a NAND-NAND CMOS circuit exists if and only if there exists a two-pattern test $<T_1,T_2>$ for the fault such that T_1 and T_2 are adjacent.

In the above result, T_1 must obviously be a 0-vertex and T_2 a 1-vertex. This result gives a method for identifying the stuck-open faults which are not robustly testable. In the example we considered earlier the test vector for detecting the stuck open fault in the pMOS transistor fed by G_4 in the circuit in Fig. 5.1 is not adjacent to any of the fifteen possible initialization vectors. This is the reason why this fault is not robustly testable.

Using Result 5.3 we first identify the stuck-open faults which are not robustly testable. Then we identify the first-level NAND gates whose outputs feed the transistors in which these stuck-open faults can occur. An extra input called the control input C is added to each of these first-level gates in order to make the circuit robustly testable. For example, the only stuck-open fault which is not robustly testable in the circuit in Fig. 5.1 is the one we considered before. Therefore we can add the control input C to gate G_4 and derive the circuit shown in Fig. 5.2.

Let us see how the stuck-open fault in the pMOS transistor fed by G_4 now becomes robustly testable. We know from before that 10100 is the only test vector possible for this fault. Suppose

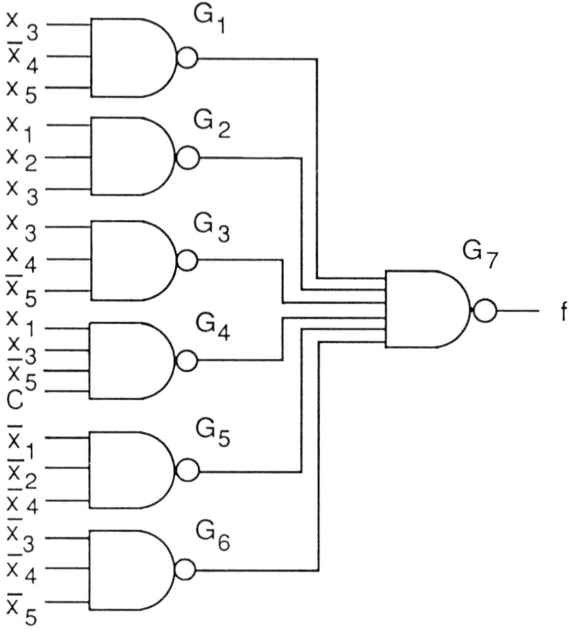

Fig. 5.2 A robustly testable realization

we feed this vector during initialization also, but make $C = 0$. Then the outputs of all the first-level NAND gates will be 1 and the output f will be 0. Next we make $C = 1$ with the other inputs fixed at 10100. The output f will become 1 in the fault-free case, but remain at 0 when the stuck-open fault is present. No spurious conduction paths can be activated in gate G_7 during the transition from the initialization vector to the test vector. Thus the stuck-open fault is robustly testable. Therefore the circuit in Fig. 5.2 is now completely robustly testable. Note that incorporation of C changes the function f. However, under normal operation C is fed 1 and the original function is realized. Thus the control input is used for test purposes only.

The above method yields circuits which are robustly testable with respect to single stuck-open faults. The robust test set derived for such circuits, in which the two vectors in any two-pattern test are adjacent, will detect all single stuck-at faults as well. If current monitoring is allowed, all single stuck-on faults will also be detected. This is owing to the fact that the test vector of a two-pattern test which detects a stuck-open fault in some transistor also detects the stuck-on fault in the dual transistor (see Section 4.1.2.1). However, it may not always be possible to implement a function with a two-level circuit. The primary reason is that in a sum-of-products expression a product term may have an arbitrary number of literals. Thus a two-level circuit based on this expression will require NAND gates with an arbitrary number of inputs. However, technological constraints limit the number of inputs that a gate can have to between 4 and 8. The problem created by this fan-in constraint is taken care of in the following result:

Result 5.4: If a two-level NAND-NAND or NOR-NOR CMOS circuit N is robustly testable for all single stuck-open faults then the circuit obtained by replacing any gate in N by a tree of primitive gates to meet the fan-in restrictions is still robustly testable for all single stuck-open faults.

The above result was later generalized to multiple faults in [KUND88b].

An application of Result 5.4 is shown in Fig. 5.3. If the fan-in of a gate is restricted to 2 then the high fan-in gates of Fig. 5.2 can be replaced by a tree of primitive gates with fan-in ≤ 2, as shown in Fig. 5.3. The replacement with the tree of gates should be such that the functionality of the gate is preserved.

Other methods were also discussed in [REDD83, REDD86] for making a complex gate robustly testable. These methods involve adding at least two extra transistors and one controllable input to

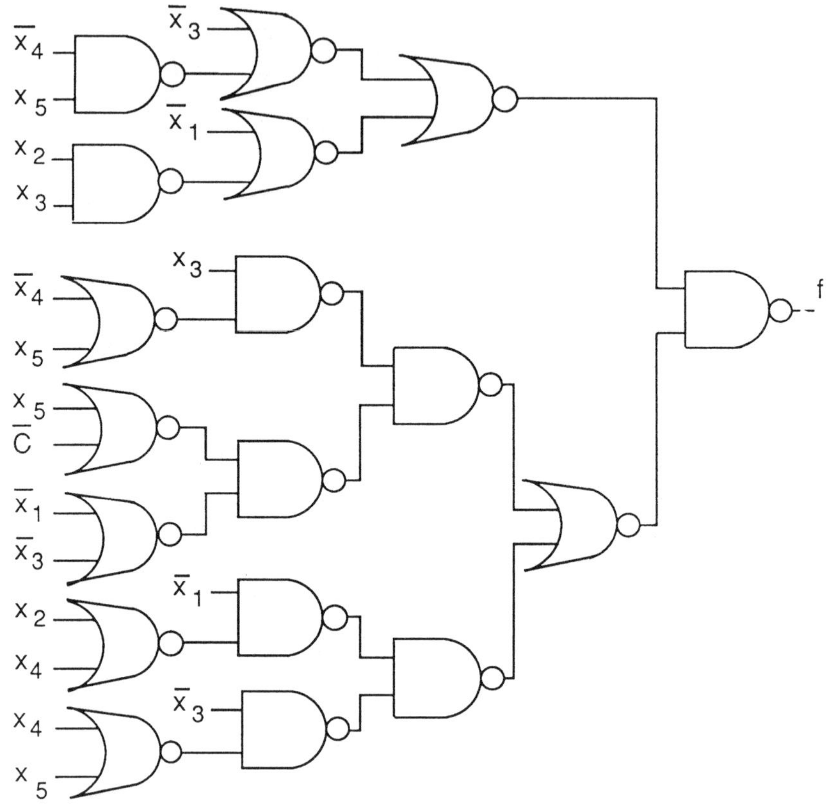

Fig. 5.3 A robustly testable circuit meeting
fan-in constraints

the complex gate.

5.1.2 The Liu-McCluskey Method

In a method for obtaining robustly testable designs by Liu
and McCluskey [LIU87], two extra inputs are used. These inputs
feed two additional transistors that are added to every CMOS gate
(except inverters) in the CMOS circuit. In addition, an inverting
buffer is inserted after every non-circuit-output CMOS gate. The

technique for modifying a CMOS gate is shown in Fig. 5.4. An extra pMOS transistor is connected in series with the pMOS network and is controlled by the control input C_p. Similarly, an extra transistor is connected in series with the nMOS network and is controlled by the control input C_n. Under normal operation, $C_p = 0$ and $C_n = 1$. These inputs are only used in the test mode.

In this design a stuck-open fault is detected by a three-pattern test $<T_1,T_2,T_3>$. Suppose there is a stuck-open fault in the pMOS network. Let T_3 be the test vector for this fault. C_p for this test vector must be 0, because only then can there be a conduction path from V_{dd} to the output node in the fault-free case. The logic value on C_n is don't-care since the nMOS network cannot conduct under T_3. Suppose that C_n is 1. The initializing vector T_1 must initialize the output node to 0. For this vector

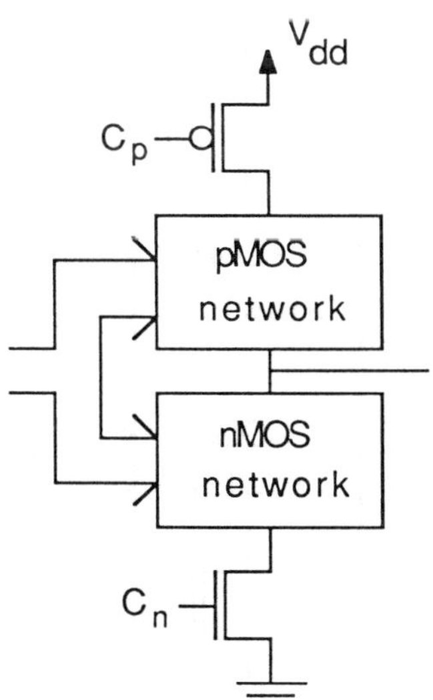

Fig. 5.4 CMOS gate modification

both C_p and C_n are 1. T_2 is derived by keeping C_p at 1 and changing the rest of the inputs in T_1 to those in T_3. This way the initialization is preserved regardless of circuit delays and the three-pattern test is robust.

Another method from [LIU87] can also be used to detect stuck-open faults with two-pattern tests instead of three-pattern tests. In this method it is assumed that an input x_i and its complement $\overline{x}_i$ can both be set to 1 or to 0. This is possible if both literals (x_i and $\overline{x}_i$) are controllable by input pins. Else, if they are controlled by scan-path bistables, a modified bistable element [LIU86] can also set both x_i and $\overline{x}_i$ to the same logic value. Under these assumptions, the output of the gate can be first initialized to 0 by feeding it an all-1 vector and making $C_p = 0$ and $C_n = 1$. Then keeping C_p and C_n fixed, the test vector for the stuck-open fault in the pMOS transistor can be applied. The resultant two-pattern test is robust. However, for the three-pattern tests derived for the stuck-open faults before, we need not assume that both an input and its complement can be set to the same logic value in the test mode. We can similarly derive three-pattern or two-pattern tests for stuck-open faults in nMOS transistors.

An additional advantage of using this design is that detection of stuck-on faults in the nMOS or pMOS network does not require current monitoring. Such faults can be detected by two-pattern tests. Consider a stuck-on fault in the pMOS network of the CMOS gate in Fig. 5.4. Let T_1 be a vector which does not create a conduction path through the pMOS network in the fault-free case, but does create a path when the stuck-on fault is present. For this vector, $C_p = 0$. If $C_n = 1$ then under fault-free operation T_1 produces a 0 at the output. However, when the fault is present, the output will attain an intermediate logic value because both networks conduct. Let T_2 be derived from T_1 by simply changing

C_n to 0. In a fault-free situation this creates a floating output node which will maintain its previous logic value 0. However, in the presence of the fault, there is a conduction path from V_{dd} to the output node and a 1 will be obtained at this node. Therefore the fault is detected. Stuck-on faults in the nMOS network can be similarly treated.

The extension of this testing technique to a CMOS circuit in which many CMOS gates are present is straightforward. The only point to remember is that an inverting buffer is placed at the output of every testable CMOS gate in the CMOS circuit which feeds any other testable CMOS gate. This is required to ensure proper propagation of the errors to the circuit outputs.

It is clear that any test which detects a stuck-open fault in the pMOS (nMOS) network of a testable gate also detects a stuck-open fault in the transistor fed by C_p (C_n) and the nMOS (pMOS) transistor of the inverting buffer that may follow the testable gate. However, to detect a stuck-on fault in either of these two extra transistors or the two transistors in the inverting buffer, we need current monitoring.

5.2 TESTABLE DESIGNS USING COMPLEX GATES

Instead of implementing a sum-of-products (product-of-sums) expression with a NAND-NAND (NOR-NOR) CMOS circuit we can also implement it with a complex gate. Robustly testable designs of complex gates have been presented in [REDD83, JHA84, JHA85, REDD86]. In theory, any function can be implemented with a single complex gate. However, in practice, it is not always possible to do so due to fan-in restrictions.

Let us consider the function $f = \overline{x}_1 x_2 + x_1 \overline{x}_2 + \overline{x}_3 \overline{x}_4 + x_3 x_4 + \overline{x}_1 x_3$, which we have also considered earlier in Chapter 2. Based on this expression we can derive the complex gate shown in Fig. 5.5. A complex gate based on the sum-of-products expression is also called an AND-OR CMOS realization. We know from Chapter 2 that the AND-OR CMOS realization of Fig. 5.5 is not robustly testable. We can similarly derive the OR-AND CMOS realization

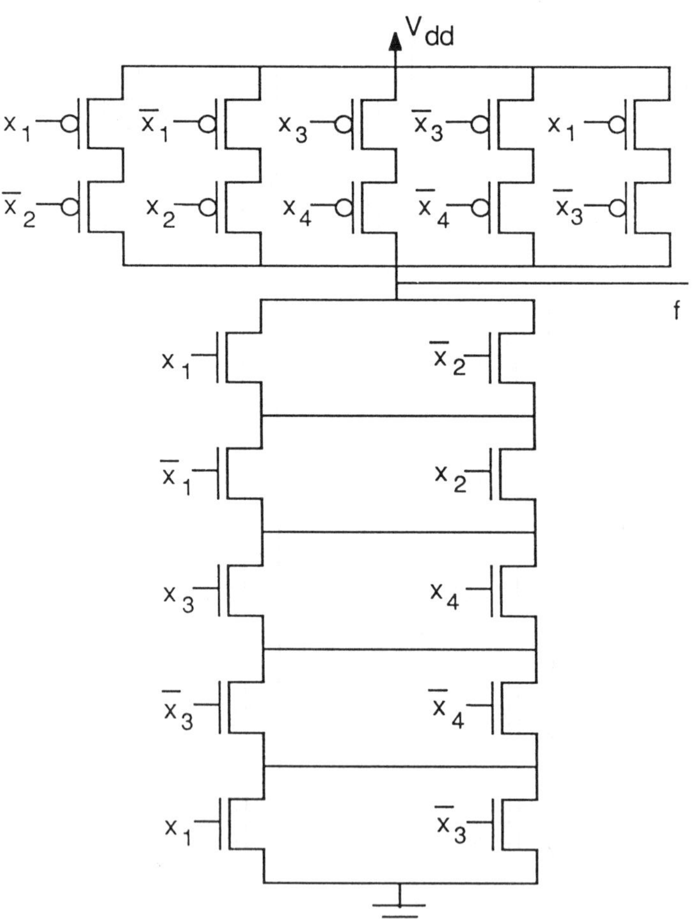

Fig. 5.5 An AND-OR CMOS realization

from the product-of-sums expression $f = (x_1+x_2+x_3+\overline{x}_4)$ $(\overline{x}_1+\overline{x}_2+x_3+\overline{x}_4)$ $(\overline{x}_1+\overline{x}_2+\overline{x}_3+x_4)$ for the same function, as shown in Fig. 5.6.

If we take the pMOS network of the OR-AND CMOS realization and attach it to the nMOS network of the AND-OR CMOS

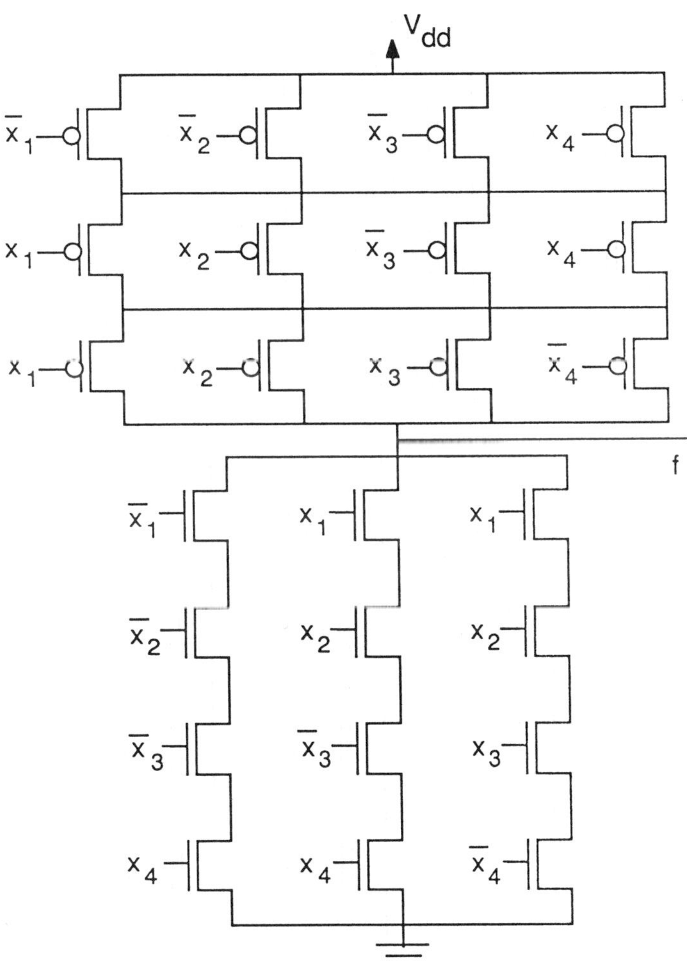

Fig. 5.6 An OR-AND CMOS realization

realization, we obtain the Hybrid CMOS realization as shown in
Fig. 5.7 [JHA84, JHA85]. This realization is referred to as the
PS-PS CMOS realization in [REDD86]. This is a perfectly valid

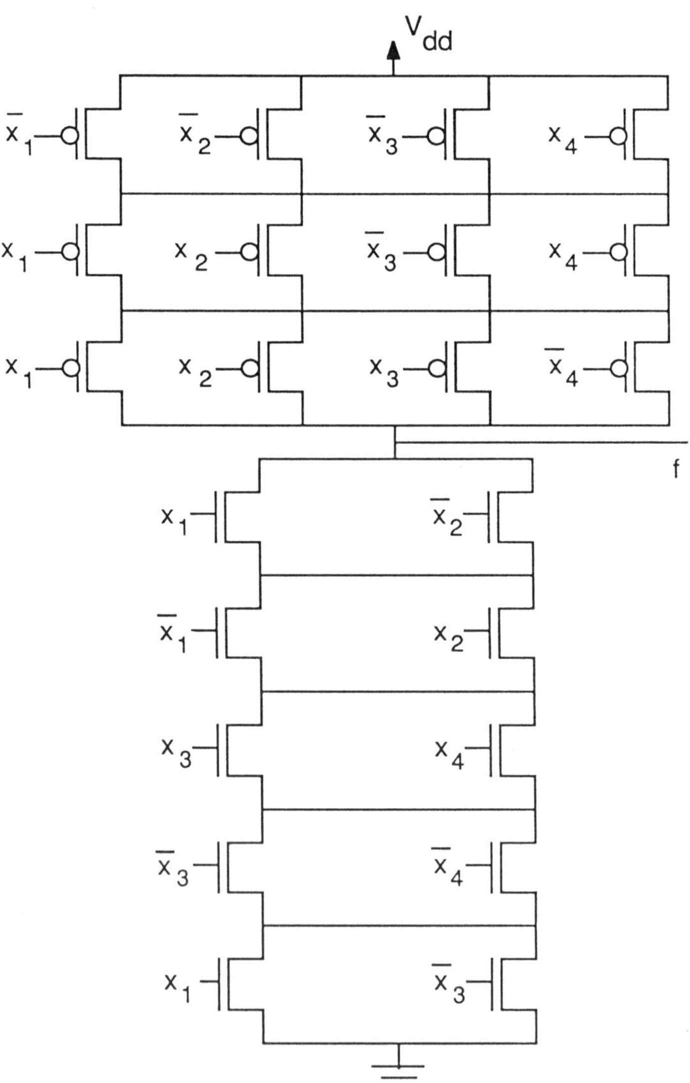

Fig. 5.7 A Hybrid CMOS realization

realization since only one of the two networks conducts for any input vector. However, the two networks are not dual in structure any more. As pointed out in [REDD86], a fourth realization can also be obtained for the function f by attaching the pMOS network of the AND-OR CMOS realization to the nMOS network of the OR-AND CMOS realization. However, such a realization does not have very good testability properties.

The following result can be derived for the Hybrid or PS-PS CMOS realization:

Result 5.5: A Hybrid CMOS realization of any function is robustly testable with respect to all single stuck-open faults.

The reason why a Hybrid CMOS realization is robustly testable lies in the way its transistors are interconnected. Both its networks basically consist of a series connection of sets of transistors which are connected in parallel. When a transistor is being tested for a stuck-open fault, all the transistors connected in parallel to it can be made non-conducting for both the initialization vector and the test vector. Thus test invalidation cannot occur.

It is possible to reduce the number of transistors in a Hybrid CMOS realization by using the law $(x + y)(x + z) = (x + yz)$ from Switching Algebra, where x, y and z can be variables or subexpressions. The resulting realization is called the Reduced Hybrid CMOS realization. One of the possible Reduced Hybrid CMOS realizations that can be obtained from the Hybrid CMOS realization of Fig. 5.7 is shown in Fig. 5.8. Any Reduced Hybrid CMOS realization is also robustly testable.

It was shown in [REDD86] that a PS-PS or Hybrid CMOS realization is also testable in the presence of charge sharing (refer to Chapter 2 for explanation on test invalidation due to charge sharing). According to another result in [JHA88] a complex gate

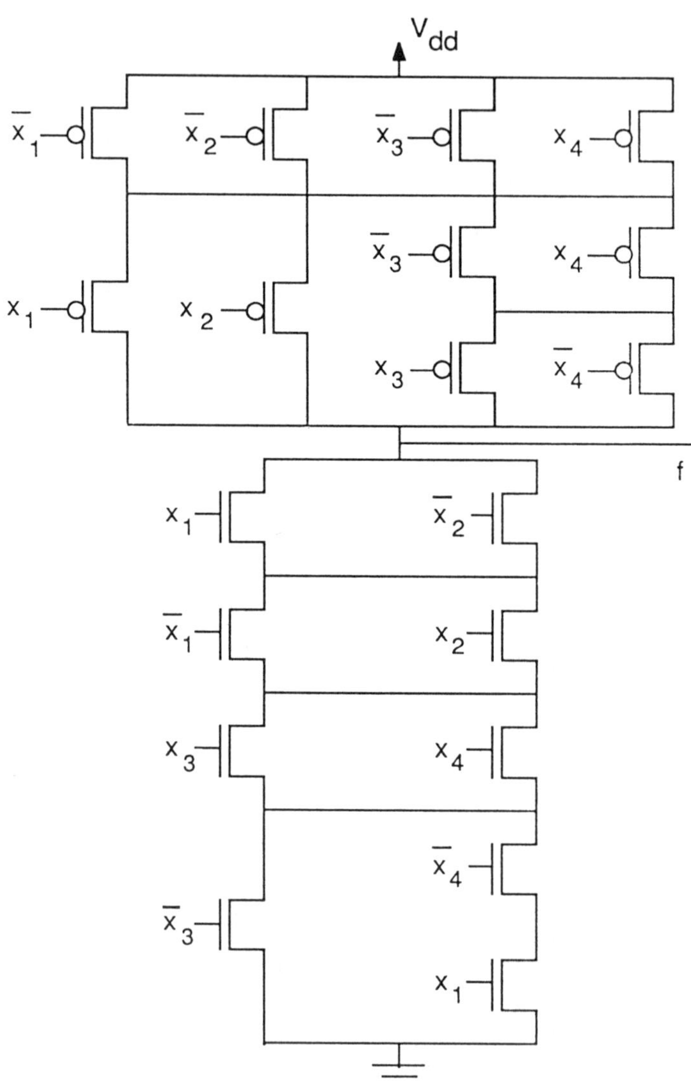

Fig. 5.8 A Reduced Hybrid CMOS realization

which is robustly testable with respect to single stuck-open faults is also robustly testable with respect to all multiple stuck-open faults. The reason is that the presence of more than one stuck-open fault can only further aid in testing. Therefore we can claim that the Hybrid CMOS realization is robustly testable with respect to all multiple stuck-open faults. The same result, of course, also holds true for the Reduced Hybrid CMOS realization.

5.3 TESTABLE DESIGN USING PARITY GATES

EX-OR or EX-NOR gates are referred to as parity gates. Using some special CMOS designs for parity gates, a CMOS circuit employing them can be made robustly testable [KUND89]. We first introduce the special designs of parity gates and describe their properties.

The gates are shown in Figs. 5.9 and 5.10. Note the resemblance of these gate structures with the structures of NAND and NOR gates. Buffering of the gates, as shown in Figs. 5.9 and 5.10, may be needed to restore signal levels if the parity gates are cascaded extensively. We next present the robust two-pattern tests that are both necessary and sufficient to detect all single stuck-

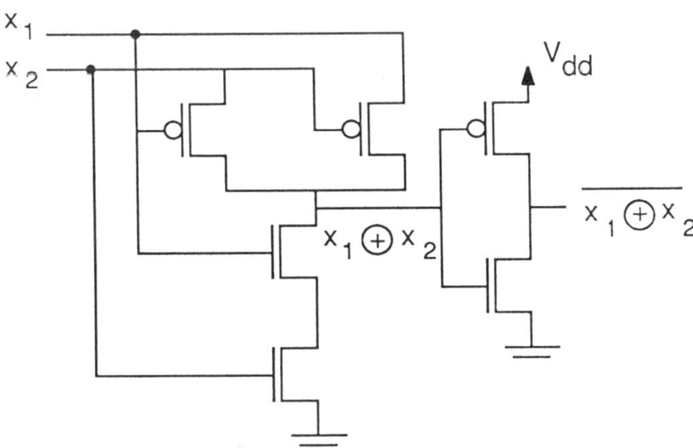

Fig. 5.9 A cross-coupled parity gate

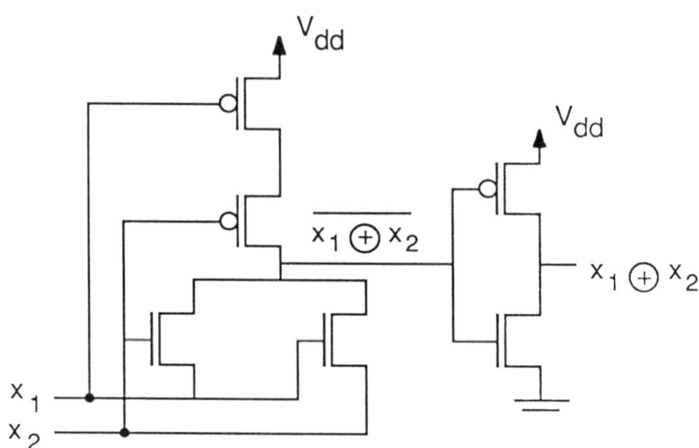

Fig. 5.10 Another cross-coupled parity gate

open and stuck-at faults. These tests will also detect all single stuck-on faults if current monitoring is done.

(1)   For the parity gate in Fig. 5.9 the test set consists of the following robust two-pattern tests: $<$00 or 11, 01$>$, $<$00 or 11, 10$>$ and $<$01 or 10, 11$>$.

(2)   For the parity gate in Fig. 5.10 the test set consists of the following robust two-pattern tests: $<$00 or 11, 01$>$, $<$00 or 11, 10$>$ and $<$01 or 10, 00$>$.

Note that even if the parity gate in Fig. 5.9 does not receive the vector 00, it can still be robustly tested by sequencing the vectors 11, 01 and 10 appropriately. One such sequence is $<$11, 01, 11, 10$>$. Similarly, even if the parity gate in Fig. 5.10 does not receive the vector 11, it can still be robustly tested.

We next explain the advantages of using parity gates for implementing any function. We know that a function can be realized in a sum-of-products form. It can also be realized in what is referred to as the linear sum-of-products form. In this form the product terms are EX-ORed. A corresponding two-level circuit can be derived which has NAND gates in the first level and a parity gate (either EX-OR or EX-NOR) at the second level. These two-level circuits are called NAND-Parity circuits. In a NAND-NAND circuit unless the product term being realized is a prime implicant the circuit becomes redundant. This, however, is not always the case with NAND-Parity circuits. This is explained through an example next.

Consider the expression $f = x_1 + \bar{x}_1 x_2$. In this expression $\bar{x}_1 x_2$ is not a prime implicant. It is merely an implicant. Whenever, $\bar{x}_1 x_2$ is 1, x_2 is also 1. Thus $\bar{x}_1 x_2$ is covered by x_2 which is a prime implicant of f. In a two-level NAND-NAND circuit which realizes the expression $x_1 + \bar{x}_1 x_2$, a stuck-at 1 fault on the line fed by $\bar{x}_1$ cannot be detected. Therefore the circuit is redundant. Now suppose that f is implemented using the linear sum-of-products expression $x_1 \oplus \bar{x}_1 x_2$. The corresponding two-level circuit would be a NAND-Parity circuit as shown in Fig. 5.11. In this circuit we have actually implemented $\bar{x}_1 \oplus \overline{\bar{x}_1 x_2}$ which is the same as $x_1 \oplus \bar{x}_1 x_2$. The following truth table can be derived for $f = x_1 \oplus \bar{x}_1 x_2$.

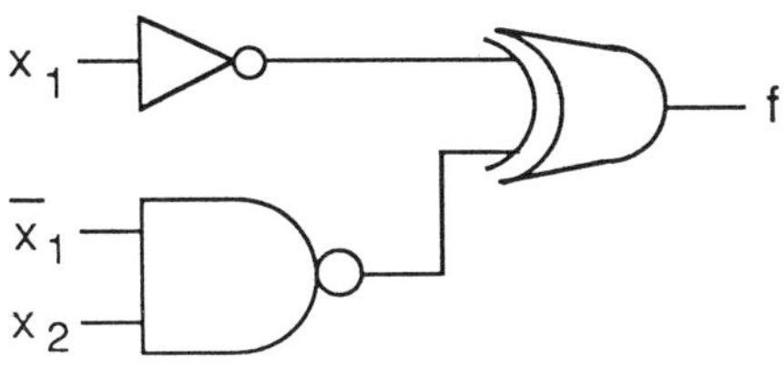

Fig. 5.11 NAND-Parity circuit for $f = x_1 \oplus \bar{x}_1 x_2$

Table 5.1 Truth table

x_1	x_2	$\overline{x}_1 x_2$	$x_1 \oplus \overline{x}_1 x_2$
0	0	0	0
0	1	1	1
1	0	0	1
1	1	0	1

The parity gate receives the following vectors: 01, 10, 11. Note that the parity gate in Fig. 5.9 needs precisely these vectors for testing purposes. Also, the NAND gate realizing $\overline{x}_1 x_2$ receives all possible input combinations. Any vector which is fed to a parity gate sensitizes a path through it. Thus the NAND-Parity circuit is irredundant. This leads to the following observation:

Observation 5.1: In a NAND-Parity circuit the product terms which are implemented need not be prime implicants of the function for the circuit to be irredundant.

We next illustrate the method for obtaining a robustly testable circuit through an example.

Example 5.1: Consider a CMOS circuit whose gate-level model is as shown in Fig. 5.12, which realizes the function $f = x_1 x_2 + \overline{x}_1 \overline{x}_2 + x_3 x_4 + \overline{x}_3 \overline{x}_4 + \overline{x}_1 \overline{x}_3$. $x_1 x_2$, $\overline{x}_1 \overline{x}_2$, $x_3 x_4$, $\overline{x}_3 \overline{x}_4$ and $\overline{x}_1 \overline{x}_3$ are the prime implicants of the function. In the CMOS circuit the stuck-open fault in the pMOS transistor of gate G_6, which is fed by gate G_5, is not robustly testable. This circuit can be transformed into a robustly testable circuit using parity gates. One way to do this is shown in Fig. 5.13. The function f in the circuit in this figure has been realized as $(x_1 x_2 + \overline{x}_1 \overline{x}_2 + x_3 x_4 + \overline{x}_1 \overline{x}_3) \oplus x_1 \overline{x}_2 \overline{x}_3 \overline{x}_4$. This, in fact, is a NAND-NAND-Parity circuit. In this expression the prime implicant $\overline{x}_3 \overline{x}_4$ has been replaced by the implicant $x_1 \overline{x}_2 \overline{x}_3 \overline{x}_4$.

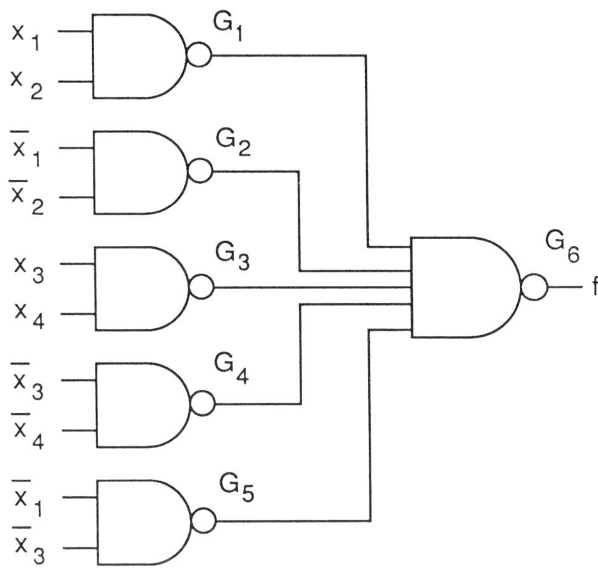

Fig. 5.12 A NAND-NAND realization

It is easy to check that the circuit in Fig. 5.13 is robustly testable.
□

An alternative way to design the circuit is to use the NAND-Parity realization discussed earlier. The function f can be realized as $x_1x_2 \oplus \overline{x}_1\overline{x}_2 \oplus \overline{x}_1x_2x_4 \oplus \overline{x}_1\overline{x}_3\overline{x}_4 \oplus \overline{x}_2\overline{x}_3\overline{x}_4 \oplus x_1\overline{x}_2x_3x_4$. The corresponding implementation is shown in Fig. 5.14, which can be easily checked to be robustly testable. Note that since $\overline{y} = 1 \oplus y$, by implementing the product terms with NAND gates instead of AND gates, we do not change the function when the number of product terms (and, hence, the number of NAND gates) is even. If the number of product terms were odd, we could still implement them with NAND gates. However, in this case the parity circuit would implement the EX-NOR function instead of the EX-OR function.

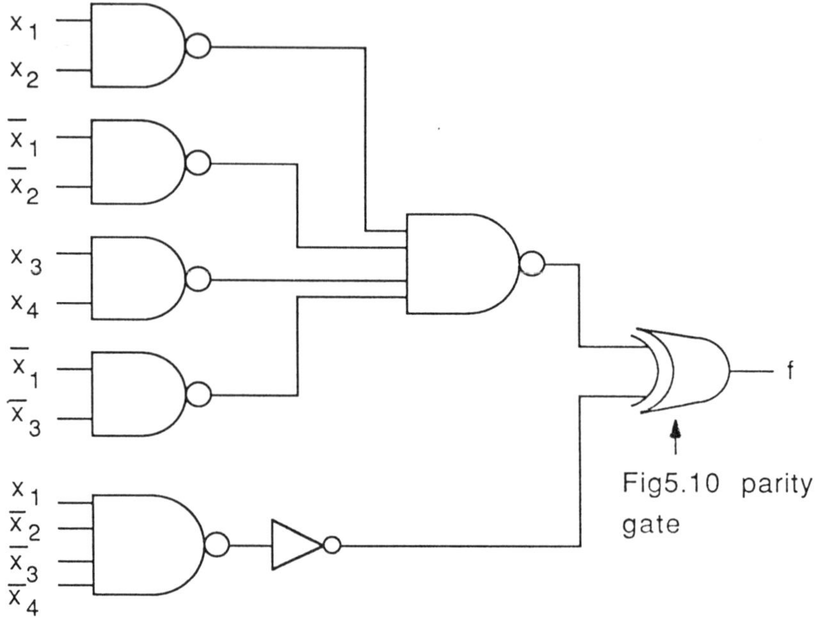

Fig. 5.13 A robustly testable realization

Minimization techniques for two-level NAND-NAND circuits is well-understood. However, little is known about two-level NAND-Parity circuit minimization techniques. Therefore, henceforth, we will follow the design approach used in Fig. 5.13. We next describe the formal method for arriving at such a design.

A switching function Q is said to imply another function P if P assumes the value 1 whenever Q does. If Q implies P then P is said to cover Q. An irredundant NAND-NAND circuit realizes a sum-of-products expression in which all product terms are prime implicants [KOHA78]. Let a function f given by $\sum_{i=1}^{m} P_i$ be realized by an irredundant NAND-NAND circuit as shown in Fig. 5.15, where P_i represents a prime implicant whose complement is

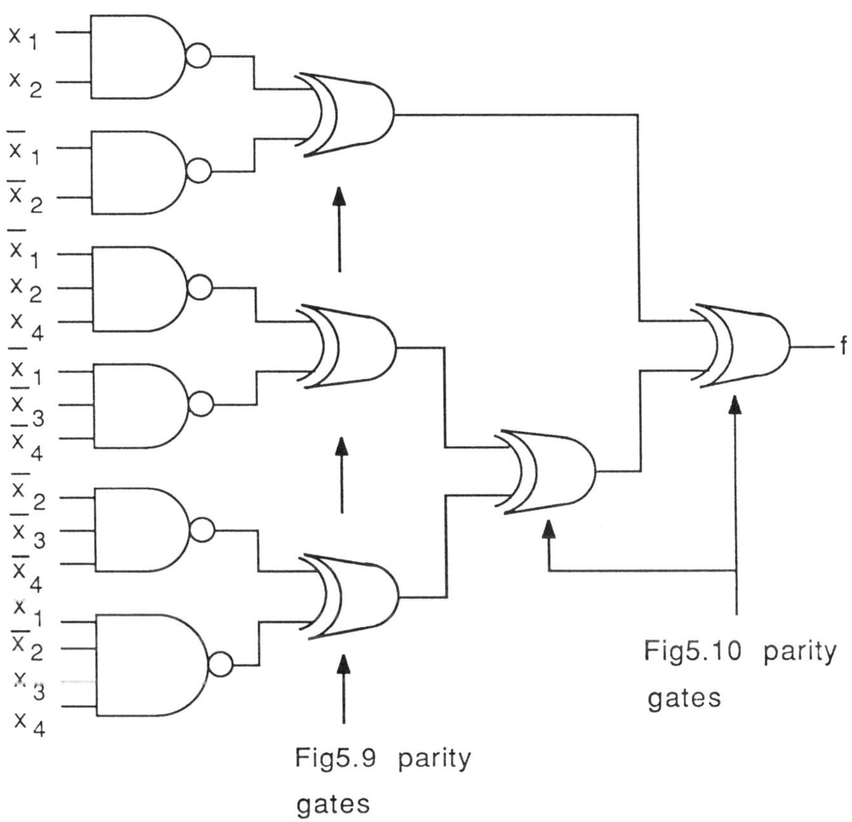

Fig. 5.14 A NAND-Parity realization

realized by gate G_i. Since the circuit in Fig. 5.15 is irredundant, there exists a test for the stuck-at 1 fault at the output of gate G_i. A test for this fault should result in a 0 at the output of G_i and a 1 at the outputs of all other first-level gates in the fault-free case. Therefore there exists a 1-vertex of the function f that implies the prime implicant P_i, but does not imply any other prime implicant P_j, $1 \leq j \leq m$, $i \neq j$. This 1-vertex is called the *distinguished vertex* of the prime implicant P_i in the expression $f = \sum_{i=1}^{m} P_i$. A prime implicant may have more than one distinguished vertex.

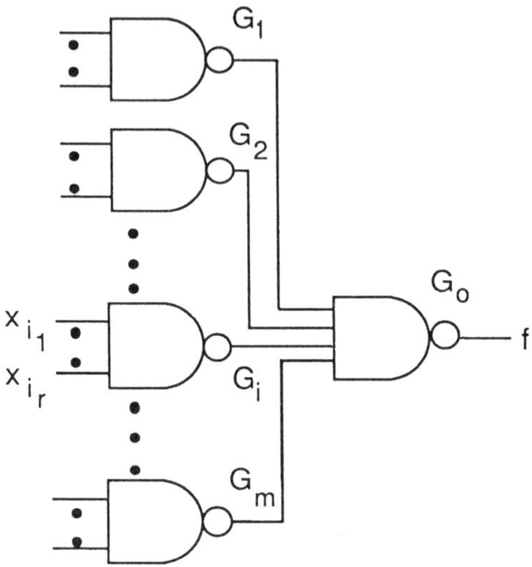

Fig. 5.15 A NAND-NAND circuit

If a prime implicant P covers a 1-vertex of the function f, which cannot be covered by any other prime implicant Q, then P is said to be an *essential prime implicant*, otherwise it is called a *non-essential prime implicant*.

Consider the circuit shown in Fig. 5.12 again. In this circuit x_1x_2, $\overline{x}_1\overline{x}_2$, x_3x_4, $\overline{x}_3\overline{x}_4$ and $\overline{x}_1\overline{x}_3$ are prime implicants of the function f. Prime implicants x_1x_2, $\overline{x}_1\overline{x}_2$, x_3x_4 and $\overline{x}_3\overline{x}_4$ are essential. However, $\overline{x}_1\overline{x}_3$ is a non-essential prime implicant of f, because every 1-vertex of f that implies $\overline{x}_1\overline{x}_3$ also implies some other prime implicant of f. In the expression $f = x_1x_2 + \overline{x}_1\overline{x}_2 + x_3x_4 + \overline{x}_3\overline{x}_4 + \overline{x}_1\overline{x}_3$, $\overline{x}_1\overline{x}_3$ has only one distinguished vertex 0101. 0101 is a 1-vertex that does not imply x_1x_2, $\overline{x}_1\overline{x}_2$, x_3x_4 or $\overline{x}_3\overline{x}_4$. It, however, implies $\overline{x}_1x_4$ which is also a prime implicant of f, though not included in the expression for f. This prevents $\overline{x}_1\overline{x}_3$ from being an essential prime implicant of f.

We are now in a position to state the following result, which is a modification of Result 5.3 [REDD83, REDD86] mentioned earlier.

Result 5.6: In a two-level irredundant NAND-NAND CMOS circuit implementing the function f, the stuck-open faults in the pMOS transistors of the output gate can be robustly tested if and only if every prime implicant whose complement is realized by a first-level NAND gate has a distinguished vertex adjacent to a 0-vertex of f.

With the help of the above result and Result 5.1 mentioned earlier, we can verify if a NAND-NAND CMOS circuit is robustly testable. To make this verification even easier, we can make use of the following result from [KUND89].

Result 5.7: Every essential prime implicant of a function f has a distinguished vertex adjacent to a 0-vertex of f.

By combining Results 5.6 and 5.7 the following results can be obtained:

Result 5.8: The stuck-open faults in the pMOS transistors of the output gate of a two-level NAND-NAND CMOS circuit driven by gates realizing the complements of essential prime implicants are robustly testable.

Result 5.9: If a stuck-open fault in the pMOS transistor of the output gate of a two-level NAND-NAND CMOS circuit is not robustly testable, then it must be driven by a gate which realizes the complement of a non-essential prime implicant.

It should be pointed out that the converse of Result 5.9 is not true. This means that if a two-level NAND-NAND CMOS circuit realizes a sum-of-products expression in which one of the prime implicants is non-essential, then it is still possible that the stuck-open fault in the pMOS transistor of the output gate fed by the gate realizing the complement of this non-essential prime implicant is robustly testable.

Based on the results given so far, the following design procedure can be suggested.

Procedure Testable_Design_1(f_1, f_2, $\cdots$, f_k)

(1)   Minimize each of the functions $f_1, f_2, \cdots, f_k$ for an irredundant two-level NAND-NAND implementation.

(2)   Check the stuck-open faults in the pMOS transistors of the k output gates for robust testability. If a stuck-open fault in some output gate does not have a robust test, then it must be fed by a gate which realizes the complement of a non-essential prime implicant. Identify using a heuristic (given later) a prime implicant whose removal prevents test invalidation, and realize its distinguished vertices only.

(3)   Use parity gate(s) to combine the realizations of the distinguished vertices found in Step 2 with the output of the NAND-NAND circuit which realizes the sum of the other prime implicants.

In Step 1 of the above procedure, first-level NAND gates which implement the complement of a prime implicant that is common to two or more functions can obviously be shared among the NAND-NAND realizations of these functions. A similar procedure can also be used for NOR-NOR realizations.

Example 5.1 given earlier demonstrates how the above procedure works. We were interested in obtaining a robustly testable realization for $f = x_1x_2 + \overline{x}_1\overline{x}_2 + x_3x_4 + \overline{x}_3\overline{x}_4 + \overline{x}_1\overline{x}_3$. Out of the five prime implicants, $\overline{x}_1\overline{x}_3$ is the only non-essential one, which prevents the stuck-open fault in the pMOS transistor fed by G_5 in the circuit in Fig. 5.12 from being robustly testable. If we remove the prime implicant $\bar{x}_3\bar{x}_4$ then the prime implicant $\bar{x}_1\bar{x}_3$ will have a distinguished vertex adjacent to a 0-vertex of f. The only distinguished vertex of $\bar{x}_3\bar{x}_4$ is $x_1\bar{x}_2\bar{x}_3\bar{x}_4$. This vertex is realized separately. Then the output of the NAND-NAND circuit which realizes the sum of the other four prime implicants is EX-ORed with the realization for $x_1\bar{x}_2\bar{x}_3\bar{x}_4$ in order to obtain a robustly testable realization of f.

In the above example, instead of replacing $\bar{x}_3\bar{x}_4$ by its distinguished vertex $x_1\bar{x}_2\bar{x}_3\bar{x}_4$, one could have replaced $\bar{x}_1\bar{x}_2$ by its distinguished vertex $\bar{x}_1\bar{x}_2x_3\bar{x}_4$. For this example, it does not really matter which one is chosen. But, in general, one would obviously want to choose the prime implicant whose removal results in the need to add the fewest distinguished vertices.

Based on the above considerations, the following heuristic is recommended for each of the non-essential prime implicants which gives rise to test invalidation.

Heuristic:

(1) Make a list of all the 0s adjacent to the non-essential prime implicant in question.

(2) For each of these 0s, find the 1-vertex of the non-essential prime implicant which is adjacent to it, and find out which other prime implicants cover the 1-vertex. Then find out how many distinguished vertices have to be generated if these prime implicants are replaced.

(3) Choose the 0 for which the number of distinguished vertices to be generated is the least and select the prime implicants accordingly.

The only issue that remains to be addressed is that of fan-in. The circuit that is obtained by the above method is a NAND-NAND-Parity circuit. Since the parity function is always realized with two-input gates given in Figs. 5.9 and 5.10, the fan-in restrictions for this part are already taken care of. To meet the fan-in restrictions for the NAND-NAND part of the circuit we can take help of Result 5.4 as before.

5.4 TESTABLE DESIGN USING SHANNON'S THEOREM

The technique for obtaining a design based on Shannon's expansion theorem, which is robustly testable with respect to all single stuck-open faults, was first presented in [KUND88a]. It was then extended to cover multiple faults in [KUND88b]. According to Shannon's expansion theorem, a function $f(x_1,x_2,..,x_i,..,x_n)$ can be expanded around a variable x_i as follows [KOHA78]:

$$f(x_1,x_2,..,x_n) = \overline{x}_i \, f(x_1,x_2,..,x_{i-1},0,x_{i+1},..,x_n)$$
$$+ \, x_i \, f(x_1,x_2,..,x_{i-1},1,x_{i+1},..,x_n).$$

This expansion theorem forms the basis for the robustly testable design. The design to be presented is robustly testable for combinations of multiple stuck-open and stuck-at faults, unlike most of the previous designs which were meant to guarantee testability with respect to single faults only. The design is robustly testable with respect to multiple path delay faults as well. We will describe testing of path delay faults in brief to make the connection apparent.

5.4.1 Path Delay Faults

Path delay is the modeled delay for a $0 \rightarrow 1$ as well as a $1 \rightarrow 0$ transition at the path input to propagate to the path output. Hence, associated with each physical path in a circuit constructed of primitive gates, there are two "logical path delays". One of the reasons for path delays to fall outside of specified range is that process variations (e.g. variations in oxide thickness, dimensions of connector lines and transistors) can increase the amount of time it takes for signals to propagate through affected regions of a VLSI chip. Delay faults caused by process variations may affect large areas of the chip (or even the entire chip). Thus tests to detect path delay faults must be able to detect them even in the presence of delay faults along other paths. Tests which can guarantee such path delay fault coverage are called *robust tests* [REDD87, PARK87]. A five-valued logic was shown to be applicable to the derivation of robust tests to detect path delay faults [REDD87]. Using this logic a test-pattern generator to derive robust tests to detect path delay faults was implemented and applied to the so called ISCAS-85 circuits [BRGL85]. These results indicate that for a large number of paths in most of the ISCAS-85 circuits, robust tests to detect path delay faults do not exist. For this reason it is important to derive methods for designing primitive gate static CMOS logic circuits in which all paths are robustly testable for delay faults.

The following example deals with path delay fault testing:

Example 5.2: Consider the circuit in Fig. 5.16. There are five different paths from primary inputs to the circuit output. Each of these has to be tested to see if a $0 \rightarrow 1$ as well as a $1 \rightarrow 0$ transition at the path input reaches the circuit output within the specified amount of time. Consider the path 1-6-8. The two-pattern test $<010, 110>$ creates a $0 \rightarrow 1$ transition at line 1. A $0 \rightarrow 1$ transition is expected at the output line 8. If the transition does not occur in the specified time interval, then the delay fault is detected. Note that this is a robust two-pattern test for detecting the delay fault. The reason is that only one bit changes from the initialization vector to the test vector. The other robust two-pattern test for the $1 \rightarrow 0$ transition at the input of this path is $<110, 010>$. Similar robust two-pattern tests can be derived for the other four paths in the circuit as well. Thus this circuit is robustly testable for all path delay faults. $\square$

The following theorem from [KUND88b] establishes a connection between testing of path delay faults and testing of multiple

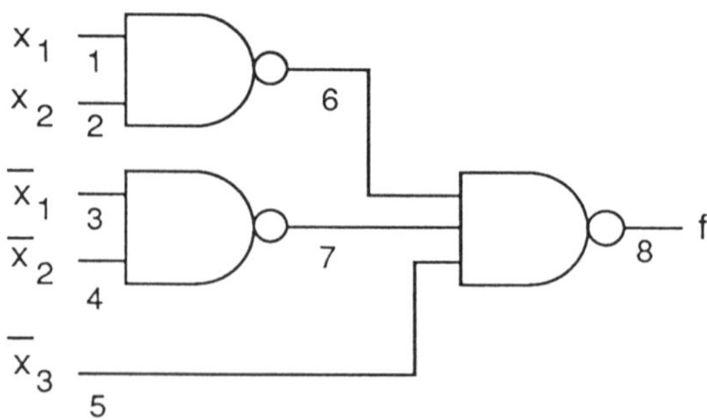

Fig. 5.16 An example circuit

stuck-open and stuck-at faults.

Theorem 5.1: If an irredundant two-level CMOS circuit is robustly testable with respect to all path delay faults, then a robust path delay fault test set exists which also detects all multiple stuck-open and stuck-at faults and their combinations.

As we saw earlier, it is very easy to derive a robust test set for all path delay faults in a two-level CMOS circuit which is robustly testable for all such faults. If there are n paths in the circuit, we need 2n two-pattern tests. A test set is obtained by appropriately sequencing these 2n two-pattern tests. Implied by Theorem 5.1 is the fact that the same test set will also detect all multiple stuck-open and stuck-at faults and their combinations. The simultaneous presence of stuck-open and stuck-at faults in the circuit does not pose any problems. It is worth keeping in mind that the concept of a robust test is needed only for a single stuck-open fault or a multiple fault at least one of whose constituents is a stuck-open fault. There is no need for this concept for a single stuck-at fault or a multiple fault all whose constituents are stuck-at faults.

It is easy to check that a test set derived as above which detects all path delay faults in a two-level circuit also detects all single stuck-at faults in it. With this observation in mind, Theorem 5.1 can be viewed as a generalization of the following well-known theorem [KOHA78].

Theorem 5.2: In an irredundant two-level circuit all multiple stuck-at faults are detected by any single stuck-at fault test set.

We next present the path delay fault testability properties of two-level circuits which implement unate functions (defined next). These results will prove to be useful when we later discuss the method for obtaining a robustly testable design for arbitrary functions.

A function f of n variables x_1, x_2,.., x_n is said to be *unate* in variable x_i if and only if $f(x_1, x_2,.., x_n)$ can be written as a sum-of-products expression in which x_i is used either in its uncomplemented or the complemented form, but not both. For example, the function $f(x_1,x_2,x_3,x_4) = x_1x_2 + x_2\overline{x}_3\overline{x}_4 + \overline{x}_1x_3 + x_1\overline{x}_4$ is unate in variables x_2 and x_4, but not in x_1 or x_3. Furthermore, since the function f is unate in the uncomplemented form of x_2, it is called *positive unate* in x_2. Similarly, f is *negative unate* in x_4. If a function is not unate in a variable x_i then x_i is called a *non-unate* or a *binate* variable of f. A function is called a unate function if and only if it is unate in all its variables.

The following theorem can be stated for two-level circuits implementing unate functions [KUND88a].

Theorem 5.3: Any irredundant two-level realization of a unate function is robustly testable with respect to all path delay faults.

This theorem is derived from the fact that in an irredundant two-level realization of a unate function, a two-pattern test $<T_1,T_2>$ can be found for detecting any path delay fault in the circuit in which T_1 and T_2 differ in only one bit. Thus in the transition from T_1 to T_2, no transients can occur.

Example 5.3: Consider the circuit in Fig. 5.17. This circuit implements the unate function $f(x_1,x_2,x_3) = x_1\overline{x}_2 + \overline{x}_2x_3$. One of the two two-pattern tests for detecting a path delay fault on the path 3-6-7 is $<001, 011>$. The two vectors differ in only one bit. Thus the two-pattern test is robust. Similar robust two-pattern tests can be found for every path in the circuit. $\square$

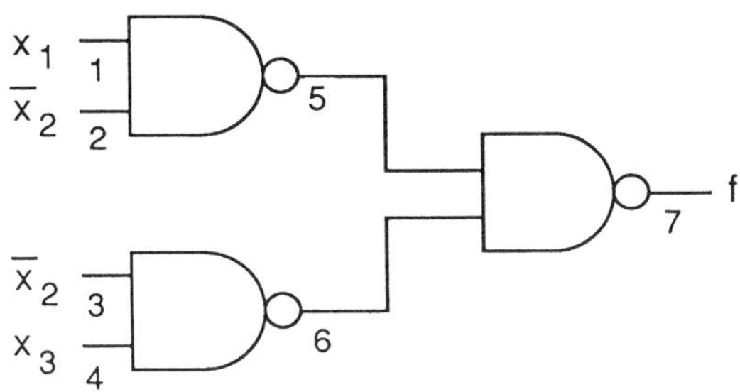

Fig. 5.17 A two-level circuit for a unate function

Note that a two-level circuit does not necessarily have to implement a unate function in order to be robustly testable with respect to all path delay faults. For example, the circuit in Fig. 5.16 does not implement a unate function, yet it is robustly testable for all path delay faults. However, robust testability is not guaranteed for two-level implementations of any arbitrary function. But Theorem 5.3 does give such a guarantee for two-level implementations of unate functions.

By combining Theorems 5.1 and 5.3 we can state the following theorem:

Theorem 5.4: A robust path delay fault test set exists for any irredundant two-level realization of a unate function which also detects all multiple stuck-open and stuck-at faults and their combinations.

The robust path delay fault test set is derived by appropriately sequencing the two-pattern tests derived for detecting all the path delay faults in the circuit.

We are now in a position to present the method for obtaining robustly testable designs for any arbitrary function.

5.4.2 Robustly Testable Design

If an irredundant two-level circuit realizing the function $f(x_1,x_2,..,x_n)$ is not robustly testable with respect to all path delay faults then f must be binate with respect to at least one of its variables. Let x_i be a binate variable of the function f. Define f_{x_i} and $f_{\overline{x}_i}$ as follows:

$$f_{x_i} = f(x_1,x_2,..,x_{i-1},1,x_{i+1},..,x_n)$$

$$f_{\overline{x}_i} = f(x_1,x_2,..,x_{i-1},0,x_{i+1},..,x_n)$$

Clearly, both f_{x_i} and $f_{\overline{x}_i}$ are independent of x_i. Using Shannon's expansion theorem [KOHA78], f can be written as

$$f(x_1, x_2,.., x_n) = x_i f_{x_i} + \overline{x}_i f_{\overline{x}_i} \tag{1}$$

$$f(x_1, x_2,.., x_n) = (x_i + f_{\overline{x}_i})\,(\overline{x}_i + f_{x_i}) \tag{2}$$

The decomposition implied by Equation (1) can be carried out by the circuit shown in Fig. 5.18, whereas the decomposition implied by Equation (2) can be implemented by the circuit in Fig. 5.19.

Lemma 5.1: Given a non-unate function f and a binate variable x_i of f, there exists a vector which can be applied to inputs $x_1, x_2, .., x_{i-1}, x_{i+1}, .., x_n$ such that $f_{x_i} = 1$ and $f_{\overline{x}_i} = 0$ ($f_{x_i} = 0$ and $f_{\overline{x}_i} = 1$).

Proof: Suppose that there does not exist a vector for which $f_{x_i} = 1$

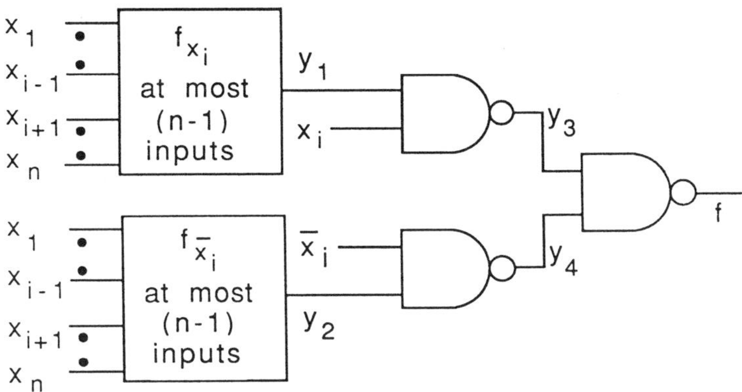

Fig. 5.18 Decomposition based on Equation (1)

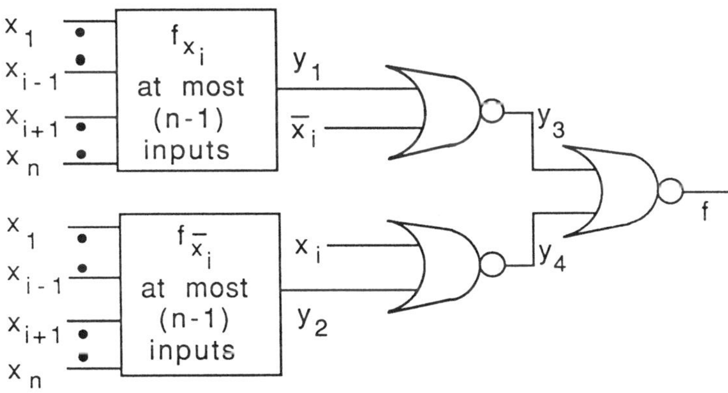

Fig. 5.19 Decomposition based on Equation (2)

and $f_{\overline{x}_i} = 0$. This means that for every vector for which $f_{x_i} = 1$, $f_{\overline{x}_i}$ also equals 1. This implies that $f_{\overline{x}_i}$ can be written as $f_{\overline{x}_i} = f_{x_i} + f'_{x_i}$ where f'_{x_i} is not equal to 0 and depends only on variables $x_1, x_2, .., x_{i-1}, x_{i+1}, .., x_n$. Therefore from Equation (1):

$$f = x_i f_{x_i} + \overline{x}_i f_{\overline{x}_i}$$

$$= x_i f_{x_i} + \overline{x}_i (f_{x_i} + f'_{x_i})$$

$$= f_{x_i} + \overline{x}_i f'_{x_i} \tag{3}$$

Equation (3) implies that f is unate in x_i, which is a contradiction. Similarly, it can be proved that there exists a vector such that $f_{x_i} = 0$ and $f_{\overline{x}_i} = 1$. $\square$

Lemma 5.1 helps us test the faults in the last two NAND-NAND (NOR-NOR) levels in the circuits in Fig. 5.18 (Fig. 5.19). Consider the circuit in Fig. 5.18. We know that a vector exists which can make $y_1 = f_{x_i} = 1$ and $y_2 = f_{\overline{x}_i} = 0$. We can apply this vector to the other $n-1$ inputs and change the logic values on the line fed by x_i to robustly detect a path delay fault in the path $x_i y_3 f$. Similarly, the fault on the path $\overline{x}_i y_4 f$ can be robustly detected. To sensitize the path $y_1 y_3 f$ ($y_2 y_4 f$) we simply need to make $x_i = 1$ ($x_i = 0$). Similar arguments apply to the circuit in Fig. 5.19.

The following theorem can now be stated [KUND88a]:

Theorem 5.5: When a non-unate function f is decomposed with respect to a binate variable x_i and implemented as in Fig. 5.18 or Fig. 5.19, then all path delay faults in the resultant circuits are robustly testable if these faults in the subcircuits realizing f_{x_i} and $f_{\overline{x}_i}$ are robustly testable.

This theorem can be extended to multiple stuck-open and stuck-at faults as follows [KUND88b]:

Theorem 5.6: When a non-unate function f is decomposed with respect to a binate variable x_i and implemented as in Fig. 5.18 or Fig. 5.19, then all multiple stuck-open and stuck-at faults and their combinations in the resultant circuits are robustly testable if these faults in the subcircuits realizing f_{x_i} and $f_{\overline{x}_i}$ are robustly testable.

Note that to verify if all multiple stuck-open and stuck-at faults and their combinations are detectable in the subcircuits in Theorem 5.6, we only need to verify if the subcircuits are robustly testable for all path delay faults (see Theorem 5.1). For example, if the subcircuit realizes a unate function then we know from Theorem 5.3 that it is robustly testable. Even if it does not realize a unate function, it may still be robustly testable. The test set which robustly detects all the multiple faults specified in Theorem 5.6 consists of the robust two-pattern tests derived for all the path delay faults in the circuit. Thus for generating 100 percent multiple stuck-open and stuck-at fault coverage, no drastic increase in test generation time is necessary.

In Theorems 5.5 and 5.6, if the subcircuits realizing f_{x_i} and $f_{\overline{x}_i}$ are not robustly testable for path delay faults then we can repeatedly apply this decomposition technique to the subcircuits until the final subcircuits are all robustly testable. It is easy to see that all 2-variable functions can be realized by robustly testable static CMOS primitive gate circuits. Hence, in at most $(n-2)$ applications of the decomposition technique and using realizations from Figs. 5.18 and 5.19, it is possible to realize a primitive gate static CMOS circuit in which all path delay faults (and, hence, all multiple stuck-open and stuck-at faults) are robustly testable. Thus a robustly testable circuit is guaranteed for any function. In practice, however, much fewer than $(n-2)$ decompositions are necessary.

Example 5.4: Consider the function $f(x_1,x_2,x_3,x_4) = x_1x_2 + \overline{x}_1\overline{x}_2 + x_3x_4 + \overline{x}_3\overline{x}_4 + \overline{x}_1\overline{x}_4$. One can verify that the two-level NAND-NAND CMOS realization of this expression is not robustly testable for all path delay faults. This function is binate in each of its variables. Choose one of them (say x_4). If we decompose the function around this variable we get

$$f_{x_4} = x_1 x_2 + \overline{x}_1 \overline{x}_2 + x_3$$

$$f_{\overline{x}_4} = x_1 x_2 + \overline{x}_1 \overline{x}_2 + \overline{x}_3 + \overline{x}_1$$

$f_{\overline{x}_4}$ can be further reduced to $(\overline{x}_1 + x_2 + \overline{x}_3)$.

The two-level NAND-NAND circuit for f_{x_4} was implemented earlier in Fig. 5.16 and found to be robustly testable for all path delay faults. For implementing $f_{\overline{x}_4}$ we only need one NAND gate. This is trivially testable for all path delay faults. Thus no further decomposition is necessary. The function can be implemented as shown in Fig. 5.20. From Theorem 5.6 and previous discussions, any test set which robustly detects all path delay faults in this circuit also robustly detects all multiple stuck-open and stuck-at faults and their combinations. $\square$

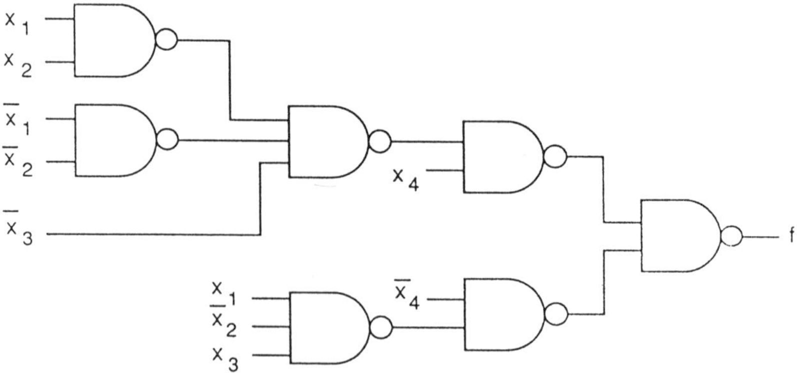

Fig. 5.20 A robustly testable realization

In the above example, x_4 was chosen as the binate variable around which decomposition was done, while we could have chosen any other variable as well. In general, choosing one variable over the other determines the maximum number of levels to which the decomposition must be carried out. This in turn determines the maximum signal propagation delay and hence the speed of the circuit. Keeping this in mind, two heuristics can be proposed for selecting the variable around which decomposition is done.

Heuristic 1: If a subfunction becomes unate then we can definitely stop further decomposition of that subfunction because from Theorem 5.3 we know that a robustly testable realization for it exists. This heuristic is geared towards quickly obtaining a unate subfunction [BRAY84]. Define the degree of a binate variable x_i to be the number of product terms of the irredundant sum-of-products expression in which x_i or $\overline{x}_i$ appears. For example, for the function $f = x_1x_2 + \overline{x}_1\overline{x}_2 + x_3x_4 + \overline{x}_3\overline{x}_4 + \overline{x}_1\overline{x}_4$, the degree of x_1 and x_4 is 3, whereas the degree of x_2 and x_3 is 2. Choose the variable with the highest degree. If more than one such variable exists, choose one arbitrarily. In the above case, either x_1 or x_4 can be selected.

Heuristic 2: Find if the two-level circuit of the function is robustly testable for all path delays. If a path in the circuit is not robustly delay testable, then identify the product term associated with it. Form a set of all such product terms. The degree of a binate variable is defined in the same way as in Heuristic 1 with the essential difference that only the product terms belonging to the set derived above are considered. The selection of the variable is then done as in Heuristic 1. For the function $f = x_1x_2 + \overline{x}_1\overline{x}_2 + x_3x_4 + \overline{x}_3\overline{x}_4 + \overline{x}_1\overline{x}_4$, the set defined above consists of only the term $\overline{x}_1\overline{x}_4$. The degree of each of the variables in $\overline{x}_1\overline{x}_4$ is 1. So either can be selected.

The design procedure is formally given next.

Procedure Testable_Design_2 (f)

(1)   Minimize the function f for a two-level design.

(2)   Check if this two-level circuit is robustly testable for all path delay faults. If yes, stop.

(3)   If the outcome in Step 2 is negative, then apply Heuristic 1 or Heuristic 2 for selecting the variable x_i around which decomposition is done.

(4)   Call Procedure Testable_Design_2 (f_{x_i}) and Procedure Testable_Design_2 $(f_{\overline{x}_i})$ to realize the circuit.

The above procedure is geared towards generating a robustly testable circuit with as few logic levels as possible. For example, if the two-level design is robustly testable then no further decomposition is done. However, in practice, one may not want to use a two-level design, because such a design may require too many gates or may not meet the fan-in restrictions or both. If reducing the fan-in of the gates is the only concern then we can replace a high fan-in gate with a tree of primitive gates realizing the same function, as we did in Sections 5.1 and 5.3. We do not sacrifice the robust testability of the circuit by doing so. In order to reduce the number of gates, we can further decompose the circuit using Heuristic 1 even if the circuit has already been found to be robustly testable. We can also share logic among the realizations of different subfunctions. By doing this we can, in general, reduce the number of gates and the fan-in of gates at the expense of some extra logic levels. We still, of course, need to check for the robust testability of the subcircuits obtained after each decomposition.

The following additional observations can be made about the design procedure:

Observation 5.2: In applying the procedure it may be advantageous to use decompositions specified by Equations (1) and (2) alternately. This would often reduce the number of logic levels in the final circuit.

Observation 5.3: Even though we have concentrated on path delay, stuck-open and stuck-at faults, it can be seen that the test set will also detect all stuck-on faults. However, we will need to monitor the current drawn by the circuit for this purpose.

REFERENCES

[BRAY84] R. K. Brayton et al., "Logic minimization algorithms for VLSI synthesis," Academic Publications, Boston, MA, 1984.

[BRGL85] Franc Brglez and H. Fujiwara, "Neutral netlist of ten combinational benchmark circuits and target translator in FORTRAN," Special session ATPG & fault simulation, in *Proc. IEEE Int. Symp. Circuits & Systems*, Kyoto, Japan, June 1985.

[CRAI87] Gary Lynn Craig, Ph.D Thesis, Department of Electrical Engineering, University of Wisconsin-Madison, 1987.

[JHA84] N. K. Jha and J. A. Abraham, "Testable CMOS logic circuits under dynamic behavior," in *Proc. Int. Conf. Computer-Aided Design*, Santa Clara, CA, pp. 131-133, Nov. 1984.

[JHA85] N. K. Jha and J. A. Abraham, "Design of testable CMOS logic circuits under arbitrary delays," *IEEE Trans. CAD*, vol. CAD-4, pp. 264-269, July 1985.

[JHA88]   N. K. Jha, "Multiple stuck-open fault detection in CMOS logic circuits," *IEEE Trans. Comput.*, vol. 37, pp. 426-432, Apr. 1988.

[KOHA78] Zvi Kohavi, Switching and Finite Automata Theory, McGraw-Hill Publishing Company, 1978.

[KUND88a] S. Kundu and S. M. Reddy, "On the design of robust testable CMOS combinational logic circuits," in *Proc. Int. Symp. Fault-Tolerant Comput.*, Tokyo, Japan, pp. 220-225, June 1988.

[KUND88b] S. Kundu, S. M. Reddy, and N. K. Jha, "On the design of robust multiple fault testable CMOS combinational logic circuits," in *Proc. Int. Conf. Computer-Aided Design*, Santa Clara, CA, pp. 240-243, Nov. 1988.

[KUND89] S. Kundu, "Design of multioutput CMOS combinational logic circuits for robust testability," *IEEE Trans. CAD*, vol. 8, Nov. 1989.

[LIU86]   D. L. Liu and E. J. McCluskey, "Design of CMOS VLSI circuits for testability," in *Proc. Custom Integrated Circuits Conf.*, Rochester, NY, pp. 421-424, May 1986.

[LIU87]   D. L. Liu and E. J. McCluskey, "Designing CMOS circuits for switch-level testability," *IEEE Design & Test*, vol. 4, pp. 42-49, Aug. 1987.

[PARK87] E. S. Park and M. R. Mercer, "Robust and non-robust tests for path delay faults in a combinational circuit," in *Proc. Int. Test Conf.*, Washington, D.C., pp. 1027-1034, Sept. 1987.

[REDD83] S. M. Reddy, M. K. Reddy, and J. G. Kuhl, "On testable design for CMOS logic circuits," in *Proc. Int. Test Conf.*, Philadelphia, PA, pp. 435-445, Oct. 1983.

[REDD86] S. M. Reddy and M. K. Reddy, "Testable realization for FET stuck-open faults in CMOS combinational logic circuits," *IEEE Trans. Comput.*, vol. C-35, pp. 742-754, Aug. 1986.

[REDD87] S. M. Reddy, C. J. Lin, and S. Patil, "An automatic test pattern generator for detection of path delay faults," in *Proc. Int. Conf. Computer-Aided Design*, Santa Clara, CA, pp. 284-287, Nov. 1987.

[SHER88] S. D. Sherlekar and P. S. Subramanian, "Conditionally robust two-pattern tests and CMOS design for testability," *IEEE Trans. CAD*, vol. 7, pp. 325-332, Mar. 1988.

ADDITIONAL READING

[CART87] J. L. Carter, V. S. Iyengar, and B. K. Rosen, "Efficient test coverage determination for delay faults", in *Proc. Int. Test Conf.*, Washington, D.C., pp. 418-427, Sept. 1987.

[CHAK89] S. Chakravarty, "On the complexity of computing tests for CMOS gates," *IEEE Trans. CAD*, vol. 8, pp. 973-980, Sept. 1989.

[KOEP86] S. Koeppe, "Modeling and simulation of delay faults in CMOS logic circuits," in *Proc. Int. Test Conf.*, Washington, D.C., pp. 530-536, Sept. 1986.

[KOEP87] S. Koeppe, "Optimal layout to avoid CMOS stuck-open faults," in *Proc. Design Automation Conf.*, Miami Beach, FL, pp. 829-835, June 1987.

[KUND88] S. Kundu, Ph.D Thesis, Department of Electrical and Computer Engineering, University of Iowa, Iowa City, Iowa 52244, May 1988.

[LIN87] C. J. Lin and S. M. Reddy, "On delay fault testing in logic circuits," *IEEE Trans. CAD*, vol. CAD-6, pp. 694-703. Sept. 1987.

[LIU87] D. L. Liu and E. J. McCluskey, "CMOS scan-path IC design for stuck-open fault testability," *IEEE J. Solid-State Circuits*, vol. SC-22, no. 5, pp. 880-885, Oct. 1987.

[SAVI86] J. Savir and W. H. McAnney, "Random pattern testability of delay faults," in *Proc. Int. Test Conf.*, Washington, D.C., pp. 263-273, Sept. 1986.

[SMIT85] G. L. Smith, "Model for delay faults based upon paths," in *Proc. Int. Test Conf.*, Philadelphia, PA, pp. 342-349, Nov. 1985.

[WAGN86] K. D. Wagner, "Delay testing of digital circuits using pseudorandom input sequences," Center for Reliable Comput., Rep. 85-12, Stanford University, revised Mar. 1986.

PROBLEMS

5.1. Prove that there is no irredundant two-level circuit, which has only three primary inputs, that has a test invalidation problem.

5.2. Find if the NOR-NOR CMOS realization of the product-of-sums expression $(x_1 + \overline{x}_2)$ $(\overline{x}_1 + x_2)$ $(x_3 + x_4)$ $(\overline{x}_3 + \overline{x}_4)$ $(\overline{x}_2 + \overline{x}_4)$ is robustly testable with respect to all single stuck-open faults. If not, modify it by adding a controllable input as in the Reddy-Reddy-Kuhl method, and derive a robust test set for the modified circuit.

5.3. Suppose that the complex gate shown in Fig. 5.5 is modified by adding two transistors with inputs C_p and C_n respectively as in the Liu-McCluskey method. Derive a test set for the modified complex gate which detects all single stuck-open and stuck-on faults. Assume that current monitoring is done to detect the stuck-on faults in the transistors fed by C_p and C_n.

5.4. Derive a Reduced Hybrid CMOS realization for the function $f = x_1x_2 + \overline{x}_1\overline{x}_2 + \overline{x}_3x_4 + x_3\overline{x}_4 + \overline{x}_1x_3$. Obtain a robust test set for this realization.

5.5. Obtain a robustly testable NAND-NAND-Parity CMOS circuit and its robust test set for the function $f = x_2\overline{x}_3x_4 + x_1x_2x_5 + x_2x_3\overline{x}_4 + x_2\overline{x}_4x_5 + \overline{x}_1\overline{x}_3\overline{x}_5 + \overline{x}_2\overline{x}_3\overline{x}_4$.

5.6. Obtain a robustly testable NOR-NOR-Parity CMOS circuit and its robust test set for the function $f = (x_1 + x_2)(\overline{x}_1 + \overline{x}_2)(x_3 + \overline{x}_4)(\overline{x}_3 + x_4)(x_1 + \overline{x}_3)$.

5.7. Find a robustly testable NOR-Parity circuit for the function $f(x_1,x_2,x_3,x_4) = \sum(1,2,14)$, where 1, 2 and 14 are the minterms of the function. Derive the robust test set for this realization.

5.8. Prove Result 5.8.

5.9. Derive a test set which robustly detects all path delay faults in the NAND-NAND circuit of Fig. 5.16.

5.10. Derive a robustly testable circuit using Shannon's expansion theorem for the function $f = \overline{x}_1x_2 + x_1\overline{x}_2 + x_3x_4 + \overline{x}_3\overline{x}_4 + \overline{x}_1x_3$. The circuit should consist of only NAND gates. Obtain a robust test set for this circuit.

5.11. Derive a robustly testable circuit made up of only NOR gates using Shannon's expansion theorem for the function $f = (x_1 + \overline{x}_2)(\overline{x}_1 + x_2)(x_3 + \overline{x}_4)(\overline{x}_3 + x_4)(x_1 + \overline{x}_3)$. Obtain a robust test set for this circuit.

Chapter 6
SELF-CHECKING CIRCUITS

With the increasing complexity of digital systems, ensuring the reliability of computations has become an important issue. A number of methods have been presented for the design of fault-tolerant systems. The first step in such a design is to detect faults. We would like to detect the faults as soon as they occur in order to prevent data contamination. An important concept which allows us to detect transient as well as permanent faults is the self-checking concept. Since transient faults have begun to play a dominant role in this era of VLSI circuits, they can no longer be ignored. As the name implies, a self-checking circuit is capable of automatically exposing its own faults. Another advantage of self-checking circuits is that software diagnostic programs can be simplified or even eliminated. This concept is becoming more and more attractive because the area overhead on a chip required to implement this concept is going down with the increase in the complexity of the chips.

6.1 CONCEPTS AND DEFINITIONS

Self-checking circuits are based on error-detecting codes. The general structure of a self-checking circuit is as shown in Fig. 6.1.

It consists of a functional circuit and a checker. The inputs and outputs of the functional circuit are encoded using a suitable code. Suppose that this circuit has p inputs and q outputs. Therefore, in all, 2^p input vectors and 2^q output vectors can occur. The sets of input and output vectors are each divided into two disjoint subsets - one consisting of codewords and the other consisting of non-codewords. The set of codewords is called the code space and the set of non-codewords is called the non-code space. Under normal fault-free operation the functional circuit receives a codeword from its input code space and produces a codeword from its output code space. The checker checks to see if the functional circuit has produced a codeword. If a non-codeword is produced then the checker gives an error indication at its outputs.

A question now arises as to what happens if there is a fault in the checker. We obviously do not want another checker to check this checker, as this will be a never-ending process. Fortunately, faults in a self-checking checker also result in an error

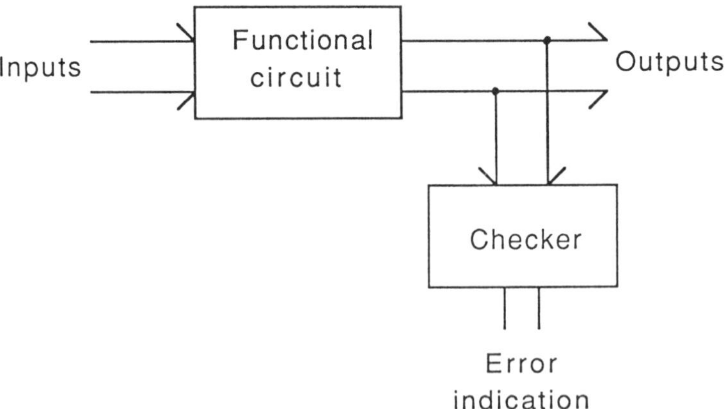

Fig. 6.1 General structure of a self-checking circuit

indication. Therefore, in response to the question: who checks the checker, we can say that it checks itself.

The concept of self-checking circuits was first proposed by Carter and Schneider [CART68]. It was later formalized by Anderson and Metze [ANDE73]. The following definitions were proposed to describe the self-checking concept.

Definition 6.1: A circuit is *fault-secure* for a set of faults F, if for any fault in F, the circuit never produces an incorrect codeword at the output for an input codeword.

Definition 6.2: A circuit is *self-testing* for a set of faults F, if for every fault in F, the circuit produces a non-codeword at the output for at least one input codeword.

It is important to understand the difference between an incorrect codeword and a non-codeword. If a circuit produces an incorrect codeword it simply means that a codeword other than the intended one is produced. On the other hand, a non-codeword is a word or vector which does not belong to the code space. If the functional circuit were to produce an incorrect codeword due to a fault then the checker would not be able to catch the error. It is precisely to prevent this situation from arising that the fault-secure property is defined. We also need to expose the faults when they occur. This is the purpose behind the definition of the self-testing property. A circuit is said to be totally self-checking (TSC) if it is both fault-secure and self-testing. For a checker, another property called the code-disjoint property is also needed. This is defined next.

Definition 6.3: A circuit is *code-disjoint* if it maps codewords (non-codewords) at its inputs to codewords (non-codewords) at its outputs.

A checker is said to be TSC if it is fault-secure, self-testing and code-disjoint. Thus if an error indication is given at the outputs of the checker then we know that there is a fault either in the functional circuit or the checker.

A self-checking checker usually has two outputs. A (0,0) or (1,1) on these outputs indicates error, whereas (0,1) or (1,0) indicates fault-free operation. The reason we need at least two outputs from the checker is that if it had only one output, the output could get stuck at the correct value and an error indication would never be given.

TSC circuits work under the following assumptions:

(1)   Faults occur one at a time;

(2)   There is a sufficient time interval between any two faults so that all the required codeword tests can be applied to the circuit.

The first assumption is required because a circuit which has been designed to be TSC with respect to all single faults from a fault set F may not be TSC if two or more faults from F are simultaneously present. The second assumption is required to make sure that a fault from F is detected before the next one occurs. Usually the set of codewords that is required to detect all the faults from F is a proper subset of the input code space of the circuit.

Given these assumptions, a TSC functional circuit always produces a non-codeword (and not an incorrect codeword) as the first erroneous output vector due to a fault. This behavior is obviously needed and is referred to as the TSC goal [SMIT78]. It is possible to design circuits which fail to be TSC, but which achieve the TSC goal under the same fault assumptions. A new concept called the strongly fault-secure (SFS) concept was presented in [SMIT78] for functional circuits. SFS circuits also meet the TSC goal. This concept is defined next.

Definition 6.4: A circuit is said to be *strongly fault-secure* with respect to a set of faults F if for every fault in F, either
(a) the circuit is self-testing and fault-secure, or
(b) the circuit is fault-secure, and if another fault from F occurs in the circuit then either property (a) or (b) is true for the fault sequence.

This definition implies that an SFS circuit continues to be fault-secure even after an undetectable fault or fault sequence is followed by another fault. Therefore it does not produce an incorrect codeword as the first erroneous output vector and thus meets the TSC goal. In fact, SFS circuits are the largest class of functional circuits which meet the TSC goal.

For checkers, too, new properties have been defined. The first one is called the strongly code-disjoint (SCD) property [NICO84].

Definition 6.5: A circuit is *strongly code-disjoint* with respect to a set of faults F if before the occurrence of any fault, the circuit is code-disjoint, and for every fault in F, either
(a) the circuit is self-testing, or
(b) the circuit always maps non-codewords at its inputs to non-codewords at its outputs, and if another fault from F occurs then either property (a) or (b) is true for the fault sequence.

An SCD checker continues to remain code-disjoint even if

undetectable faults or fault sequences are present in it. The fault-secure property was deemed to be unnecessary for the SCD checker. In fact, it has been pointed out that the fault-secure property is unnecessary for TSC checkers as well [TAMI84]. The reason is that it does not matter if the response from the checker changes from (0,1) to (1,0) or *vice versa* in the presence of the fault as long as the fault gets detected later or as long as the checker keeps mapping non-codewords at its inputs to non-codewords at its outputs. This argument is also valid for a self-checking system in which all the checkers are only self-testing and code-disjoint, but not fault-secure. This is explained next.

Consider the self-checking system shown in Fig. 6.2. In this system there are two functional circuits, the outputs of the first feeding the next. The outputs of these functional circuits are

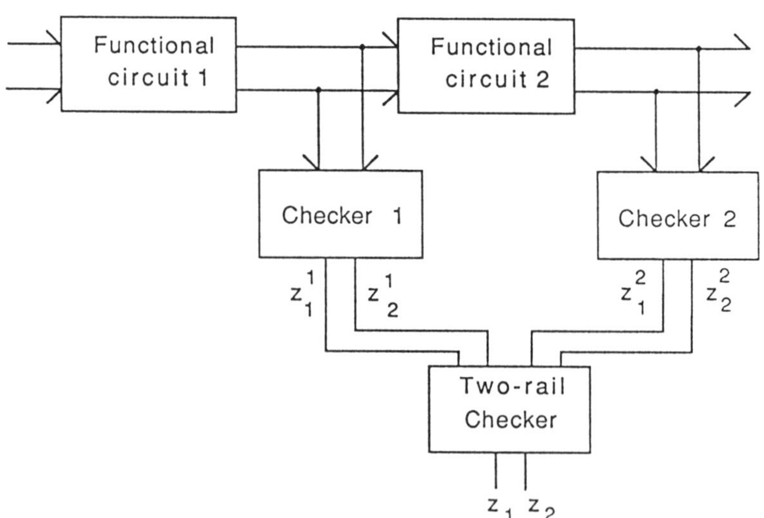

Fig. 6.2 A self-checking system

monitored by checker 1 and checker 2 respectively. Normally, the outputs of these two checkers would not be directly observable, but they would be fed to a two-rail checker, whose outputs (z_1, z_2) would be observable. A two-rail checker simply verifies if its two input rails carry complementary logic values. In the system in Fig. 6.2, (z_1^1, z_1^2) forms one rail and (z_2^1, z_2^2) forms the other. Suppose that all the checkers in the system are just self-testing and code-disjoint, but not fault-secure. Let a fault ϕ_1 in checker 1 change $(0,1)$ at (z_1^1, z_2^1) to $(1,0)$ or *vice versa* for some input codewords. However, since the checker is self-testing, there exists at least one input codeword which will expose the fault, in other words, result in $(0,0)$ or $(1,1)$ at its two outputs. Since the two-rail checker is also code-disjoint, the error will be propagated to (z_1, z_2). Thus, before a second fault occurs in the system, the fault ϕ_1 will be detected. Therefore the fault-secure property is not needed. The problem only arises when undetectable faults are allowed to be present in the system, as in the case of SCD checkers. When the two outputs of such checkers are not directly observable, they also need to be imparted the fault-secure property, as explained next.

Suppose that both checker 1 and checker 2 in the system in Fig. 6.2 are SCD. Consider a fault ϕ_1 in checker 1 with respect to which the checker is neither self-testing nor fault-secure (such a fault is allowed in an SCD checker). Let us examine the following scenario. Suppose that in the presence of ϕ_1, checker 1 produces a $(0,1)$ at (z_1^1, z_2^1) for only one input codeword and $(1,0)$ for all the others. Thus, either the codeword $(z_1^1, z_2^1, z_1^2, z_2^2) = 0101$ or 0110, but not both, will be available at the inputs of the two-rail checker. Therefore, in all, only three codewords can be fed to the two-rail checker, the other two codewords being 1001 and 1010. We will see in a later section that the two-rail checker in Fig. 6.2 requires all the four codewords 0101, 0110, 1001 and 1010 for self-checking purposes. Therefore, a subsequent fault ϕ_2 inside the two-rail checker, which was otherwise detectable, may no longer be detected. Next if a non-codeword occurs at the outputs of the functional circuit 1, it will map to a (0,0) or (1,1) at (z_1^1, z_2^1) due to the SCD property of checker 1. However, this may result in a codeword (either (0,1) or (1,0)) at (z_1, z_2) due to the presence of undetected fault ϕ_2 in the two-rail checker. This means that an erroneous output vector from a functional circuit has been allowed to go undetected, which defeats the purpose of a self-checking system.

The above arguments imply that the SCD property is not enough for checker 1 and checker 2. They should also have the fault-secure property. This led to the introduction of a new property called the strongly self-checking (SSC) property, which combines the SFS and SCD properties [JHA88, JHA89a].

Definition 6.6: A circuit is *strongly self-checking* with respect to a set of faults F if before the occurrence of any fault, the circuit is code-disjoint, and for every fault in F, either
(a) the circuit is self-testing and fault-secure, or
(b) the circuit is fault-secure and always maps non-codewords at the inputs to non-codewords at the outputs, and if another fault from F occurs then either property (a) or (b) is true for the fault sequence.

The SSC property can actually be used for both functional circuits and checkers. If it is used for functional circuits, then it is possible to cascade such SSC functional circuits and only check the outputs of the final functional circuit with a checker. The reason is that a non-codeword at the outputs of some internal functional circuit will propagate through to the final functional circuit outputs. On the other hand, if the functional circuits are only totally self-checking or strongly fault-secure, then a checker would be needed at the inputs and outputs of each of the functional circuits in the cascade. If the SSC property is used for a checker, then, strictly speaking, one can argue that the fault-secure property is not needed in part (a) of Definition 6.6, but only in part (b). The reason is that if the checker is self-testing with respect to the fault or fault sequence, then this fault or fault sequence will get detected before another fault can occur. Only if the fault or fault sequence is undetectable is the fault-secure property needed. However, in practice, it is extremely difficult to design checkers which are fault-secure with respect to some selected faults or fault sequences, but not with respect to others. The reason is that due to the large number of faults and fault sequences possible, it is not easy to investigate every fault or fault sequence to determine if the checker should be fault-secure with respect to it or not. Thus we do not lose anything by imposing the stricter requirement that the fault-secure property should be included for the checker in both part (a) and (b) in Definition 6.6, not just in part (b). Note that for a functional circuit, it is important to have the fault-secure property in both part (a) and part (b) of the definition.

As we will see later, in CMOS implementations of self-checking checkers the presence of undetectable stuck-on faults can not be avoided. Thus it is not possible to simply make the implementations self-testing and code-disjoint. For such implementations the strongly self-checking property is useful.

6.2 ERROR-DETECTING CODES

Error-detecting codes are widely employed in fault-tolerant computer systems. In particular, the inputs and outputs of a self-checking circuit are assumed to be encoded with a suitable error-detecting code. Many error-detecting codes are known. The choice of the code depends on what errors are most likely.

The types of errors can be broadly categorized as symmetric, asymmetric or unidirectional. A symmetric error involves an error in one or more bits of the data word, where an affected bit can change from 0 to 1 or 1 to 0. When only either 0 to 1 or 1 to 0 transitions occur in any data word and the type of transition is known *a priori*, the error is termed asymmetric. A unidirectional error is said to occur if all transitions are either 0 to 1 or 1 to 0, but not both, in any given data word. However, the type of transitions is not fixed *a priori*. In other words, one data word may have a unidirectional error of the 0 to 1 type while another data word may have a unidirectional error of the 1 to 0 type.

The different types of codes can be categorized as systematic, separable or non-systematic. Both systematic and separable codes consist of codewords in which a check symbol is appended to the information symbol. Suppose the number of bits in the information symbol is k and the number of bits in the check symbol is r. In a systematic code each one of the 2^k information symbols appears as a part of some codeword. In a separable code only $s < 2^k$ information symbols appear as a part of some codeword. However, the word "separable" has also been used at times in the literature to denote systematic codes.

In the case of non-systematic codes the information bits and the checkbits can not be separately identified. The advantage of these codes is that they are less redundant (the difference in the number of bits in an encoded data word and a non-encoded data word is defined to be the redundancy of the code) than systematic codes. The advantage of systematic and separable codes is that the encoding/decoding and data manipulation can be done in parallel.

Different codes have been proposed for different applications.

We discuss the most popular ones next.

Parity code. Let $X = (x_1, x_2, .., x_n)$ denote an arbitrary data word (information symbol) that has to be encoded. The check symbol for a parity code consists of only one bit x_{n+1}, which is derived by taking the modulo 2 sum of all the bits in X.

For example, if $X = 0110$ then the checkbit $x_5 = 0 \oplus 1 \oplus 1 \oplus 0 = 0$. Thus 01100 is a codeword in the parity code.

A codeword derived by the above scheme will always have an even number of 1s. Therefore it is also referred to as the even-parity scheme. The checkbit x_{n+1} can also be derived by taking the complement of the modulo 2 sum of all the bits in X. Such a code will consist of codewords with an odd number of 1s. Therefore this scheme is referred to as the odd-parity scheme. A symmetric error in an odd number of bits in either scheme will change a codeword into a non-codeword and will hence be detected. However, since it can not detect double errors, the parity code is said to be single error-detecting.

Two-rail code. As the name implies, a two-rail code consists of two rails $X = (x_1, x_2, .., x_n)$ and $Y = (y_1, y_2, .., y_n)$ where $y_i = \overline{x}_i$, $1 \leq i \leq n$.

Any symmetric error which does not only affect the corresponding bits in X and Y can be detected by this code. For example, let $X = 1000$ and $Y = 0111$. Let the symmetric error be such that X changes to $X' = 0100$ and Y changes to $Y' = 1111$. This error is detected because although the first bit of both the rails are affected, the second bit of only X is affected but not of Y.

A two-rail code can also detect any unidirectional error. However, more efficient codes can be used if protection against only unidirectional errors is required.

Residue code. The residue Q of a positive integer P with respect to a divisor m is given by

$$Q = P \text{ modulo } m.$$

The residue code is formed by concatenating the information bits which represent P and the checkbits which represent Q. Thus it is a systematic code. The divisor m is said to be the checkbase of the code. The number of checkbits $r = \lceil \log_2 m \rceil$.

As an example, let $P = 13$, which can be represented in binary form as 1101, and let $m = 3$. Then $Q = 13$ modulo $3 = 1$. Since the number of checkbits $r = \lceil \log_2 3 \rceil = 2$, the residue Q can be represented as 01. Thus the codeword will take the form 110101.

Residue codes find applications in arithmetic units [AVIZ71]. Let P_1 and P_2 be two positive integers whose residues are Q_1 and Q_2 respectively. Then

$$(P_1 \pm P_2) \bmod m = (Q_1 \pm Q_2) \bmod m$$

$$(P_1 P_2) \bmod m = (Q_1 Q_2) \bmod m$$

Thus the residue of the result of the arithmetic operation on the two integers can be compared with the result of the operation in modulo m on their residues to detect errors.

If the checkbase m is of the form $(2^b - 1)$, where $b > 1$, then the code is called a low-cost residue code. The name is derived from the fact that circuits based on these codes are easier to implement.

It can be shown that if m is any odd number greater than 1 then the residue code can detect any single bit error.

6.2.1 Codes for Detecting All Unidirectional Errors

It has been reported that unidirectional errors are the most dominant type of errors in VLSI circuits [PRAD80a]. Codes which can detect all possible unidirectional errors are called all-unidirectional error-detecting (AUED) codes.

Let X and Y be two vectors. Define $N(X,Y)$ to be the number of bits in which X has 1 and Y has 0. For example, if X $= 1010$ and Y $= 0111$ then $N(X,Y) = 1$ and $N(Y,X) = 2$. X and Y are said to be unordered if $N(X,Y) \geq 1$ and $N(Y,X) \geq 1$, else they are said to be ordered. Clearly, if two vectors are unordered then no unidirectional error can change one vector into the other. On the other hand if two vectors are ordered one can change into the other due to a unidirectional error. We can generalize these results and say that a code C is AUED if and only if every pair of codewords in C is unordered [PRAD80b, BOSE82]. Many AUED codes have been devised which satisfy this condition. We describe some of them next.

m-out-of-n code. The m-out-of-n code consists of codewords with n bits in which m are 1s. Thus there are $\binom{n}{m} =$ $n!/(n-m)!m!$ codewords in the code. The number of 1s in a codeword is said to be the weight of the codeword. Since all codewords have the same weight, this code is also called a constant-weight code.

The number $\binom{n}{m}$ is maximized when $m = \lfloor n/2 \rfloor$ or $\lceil n/2 \rceil$. Thus $\lfloor n/2 \rfloor$-out-of-n or $\lceil n/2 \rceil$-out-of-n codes are the most efficient codes in this class. In fact they have been shown to be optimal among all AUED codes [FRIE62]. However, since these codes are non-systematic in nature, encoding/decoding and data manipulation can not be done in parallel.

Berger code. This code is a systematic AUED code [BERG61]. There are two methods for obtaining this code. In the first method the check symbol is just the binary representation of the number of 0s in the information symbol. In the second method the check symbol is obtained by taking the complement of every bit in the binary representation of the number of 1s in the information symbol. When there are k bits in the information symbol the count of 0s or 1s can only vary from 0 to k. Thus the number of checkbits r in the check symbol is given by $r = \lceil \log_2(k+1) \rceil$. If $k = 2^r - 1$, the code is called a maximal-length Berger code, else it is called a non-maximal-length Berger code. For a maximal-length Berger code the two methods of encoding give the same results. However, for a non-maximal-length Berger code the two methods give different check symbols for any information symbol.

Consider a maximal-length Berger code with $k = 3$ and $r = 2$. Table 6.1 shows the results of the encoding by either method. A codeword can be formed by concatenating the check symbol to the corresponding information symbol. We can check to see that every pair of codewords is unordered. Thus the code is AUED.

It has been shown in [FRIE62] that the Berger code is optimal among all systematic AUED codes. Due to its easy encod-

Table 6.1 Encoding example

Information symbol	Check symbol
0 0 0	1 1
0 0 1	1 0
0 1 0	1 0
0 1 1	0 1
1 0 0	1 0
1 0 1	0 1
1 1 0	0 1
1 1 1	0 0

ing and decoding, this code has found wide popularity.

Smith code. This code is an optimal separable AUED code [SMIT84]. Separable codes are used when not all the 2^k possible information symbols of k bits have to be encoded.

Let X and Y be two vectors. X is said to cover Y if $N(X,Y) \geq 1$ and $N(Y,X) = 0$, using the previously developed notation. In Smith's method the set S of information symbols that have to be encoded are first sorted into a list such that for two information symbols X_i and X_j from S, if X_j covers X_i then X_i precedes X_j in the list. Then this sorted information list is broken up into a minimum number of disjoint groups containing contiguous members of the list such that every pair of information symbols in any group is unordered. Let there be m such groups. Then $r = \lceil \log_2 m \rceil$ bits are required for each check symbol. The check symbol assignment is done as follows. First, we start with a set T of m different r-tuples and sort them such that if $Y_i, Y_j \in T$ and Y_i covers Y_j then Y_i precedes Y_j in the list. This is called the check list. Then the ith r-tuple from the check list is assigned to all members of the ith group in the information list. In [SMIT84] it is shown how a sorted information list can be formed so that an optimal separable code is obtained.

Suppose, for example, the set of information symbols that have to be encoded is {0000, 0001, 0101, 1000, 1011, 1100, 1101}. The information list consisting of four groups can be formed from S as follows: {(0000), (0001, 1000), (0101, 1011, 1100), (1101)}. Therefore $m = 4$ and $r = 2$. One of the possible sorted check lists is {11, 01, 10, 00}. Thus the separable code can be derived as {000011, 000101, 100001, 010110, 101110, 110010, 110100}. The first four bits of each codeword in this code are information bits and the last two bits are checkbits. Note that the Berger code would have required three checkbits for the above example.

When the set S contains thousands or millions of information symbols it may become cumbersome to obtain the Smith code. In such cases a separable code proposed in [JHA89b] may be employed. In this encoding scheme the information symbols from the set S are put into different groups based on their zero-count. If there are m groups then $r = \lceil \log_2 m \rceil$ checkbits are required. The check list consists of m r-tuples as before with the r-tuples arranged in the order of decreasing value starting from the r-tuple with maximum value. In other words, if the decimal equivalent of an r-tuple Y_i is greater than the decimal equivalent of an r-tuple Y_j then Y_i precedes Y_j in the list.

Consider the same set S of information symbols that we considered earlier: {0000, 0001, 0101, 1000, 1011, 1100, 1101}. Four groups can be formed from S based on the zero-count as follows: {(0000), (0001, 1000), (0101, 1100), (1011, 1101)}. Therefore $m = 4$ and $r = 2$. The check list is {11, 10, 01, 00}. Thus the code is

$\{000011, 000110, 100010, 010101, 110001, 101100, 110100\}$.

6.2.2 t-Unidirectional Error-Detecting Codes

In VLSI circuits the data word may consist of a large number of bits. Thus it may be highly unlikely that more than t bits will have a unidirectional error. The value of t may depend on the bit-organization, layout, etc. In such cases we can use t-unidirectional error-detecting (t-UED) codes. These codes detect a unidirectional error in t or fewer bits. The t-UED codes have less redundancy than the AUED codes and hence are preferable when the AUED capability is not required. Many t-UED codes have been presented in literature [BORD82, DONG82, BOSE85, JIIA87a, JHA89b], some of which we will discuss in this section. Another related class of codes are the t-burst unidirectional error-detecting (t-BUED) codes. These codes detect a unidirectional error in t adjacent bits. Such codes have been presented in [BOSE86, BLAU88].

In order to present the conditions that a t-UED code must satisfy, we first need to understand the concept of Hamming distance. The Hamming distance between two vectors X and Y, denoted as $d(X,Y)$, is simply the number of bits they differ in. Therefore $d(X,Y) = N(X,Y) + N(Y,X)$. For example, if $X = 01010$ and $Y = 11001$ then $N(X,Y) = 1$, $N(Y,X) = 2$ and $d(X,Y) = 3$. A code C is t-UED if and only if for all $X,Y \in C$, either X and Y are unordered or $d(X,Y) \geq t+1$ when one covers the other [BOSE85]. If two codewords are unordered then no unidirectional error can change one into the other. If they are ordered then it takes a unidirectional error in at least t+1 bits to change one into the other. Thus the code is t-UED if the above conditions are satisfied. Bose and Lin [BOSE85] have also shown that the conditions that a t-asymmetric error-detecting code has to satisfy are the same as those given above for t-UED codes.

Borden code. Borden presented a non-systematic t-UED code in [BORD82]. He showed that a code consisting of all length n words whose weight is congruent to $\lfloor n/2 \rfloor$ mod (t+1) is t-asymmetric error-detecting. Thus the codes are also t-UED.

Let n = 11 and t = 3. Then since $\lfloor n/2 \rfloor$ mod (t+1) = 1, the set C_1 consisting of 1-out-of-11 codewords will belong to the code. The weights of the other codewords in the code can be calculated to be 1 + (t+1) = 5 and 1 + 2(t+1) = 9. Thus the set C_2 consisting of 5-out-of-11 codewords and the set C_3 consisting of 9-out-of-11 codewords also belong to the 3-UED code. The only way a unidirectional error can change one codeword into another is by changing the codeword from set C_i, $1 \leq i \leq 3$, to a codeword from set C_j, $j \neq i$. However, this would require a unidirectional error in at least four bits. Thus the code is indeed 3-UED.

The Borden code is also optimal among all t-UED codes. However, because of its non-systematic nature it may not be suitable for some applications.

Bose-Lin codes. Bose and Lin presented many systematic t-UED codes in [BOSE85]. These codes require a fixed number of checkbits independent of the number of information bits. Let k and r denote the number of information bits and checkbits respectively. For any systematic t-UED code it is assumed that $k \geq 2^r$, otherwise Berger code could be employed to detect all unidirectional errors.

The parity code, which requires a single checkbit, detects any single-bit symmetric error. Therefore it also detects any single-bit unidirectional error. Thus the parity code is also an optimal 1-UED code.

Let k_0 denote the number of 0s in the information symbol. Then the check symbol CS can be derived for a 2-UED and 3-UED code as $CS = k_0 \bmod 4$ and $CS = k_0 \bmod 8$ respectively. Thus r $= 2$ for the 2-UED code and $r = 3$ for the 3-UED code. These codes are also optimal. With $r = 4$ the check symbol can be derived as $CS = (k_0 \bmod 8) + 4$ with the resultant code being 6-UED. This code is optimal if $k \geq 20$.

For $r \geq 5$, Bose and Lin gave two methods. In Method 1 the check symbol is derived as $CS = (k_0 \bmod 2^{r-1}) + 2^{r-2}$. This code can detect a unidirectional error in up to $2^{r-2} + r - 2$ bits. Method 2 is slightly more involved. Let $CS' \equiv k_0 \bmod (6 \times 2^{r-4})$. Therefore CS' is $(r-1)$ bits long. The 3 most significant bits of CS' can only be from the set $\{000, 001, 010, 011, 100, 101\}$. Define a one-to-one function f from this set to the vectors in the 2-out-of-4 code. For example, $f(000) = 0011$, $f(001) = 0101$, $f(010) = 0110$, $f(011) = 1001$, $f(100) = 1010$, and $f(101) = 1100$. The concatenation of these 2-out-of-4 vectors to the least significant $r-4$ bits of CS' gives the check symbol CS. For this code, $t = 5 \times 2^{r-4} + r-4$. When $r = 5$, both methods yield an 11-UED code. However, for $r \geq 6$, Method 2 is superior since $5 \times 2^{r-4} + r-4 > 2^{r-2} + r-2$.

Let us consider an example of encoding with Method 2. Suppose that $k = 47$ and $r = 5$. Then the zero-count k_0 can vary from 0 to 47. let $k_0 = 47$. $CS' \equiv 47 \bmod (6 \times 2^{5-4}) = 11$ (1011 in binary). Since $f(101) = 1100$ the corresponding $CS = 11001$. There are 12 distinct check symbols that can be derived in this case for distinct values of the zero-count: $\{11001, 11000, 10101, 10100, 10011, 10010, 01101, 01100, 01011, 01010, 00111, 00110\}$. Since there are 48 different values of the zero-count possible, each of these check symbols obviously has to be used more than once. For example, for $k_0 = 11$, 23 and 35 in addition to $k_0 = 47$,

CS = 11001.

6.2.3 t-Burst Unidirectional Error-Detecting Codes

The t-burst unidirectional error-detecting (t-BUED) codes detect any unidirectional error in t or fewer adjacent bits. Any t-UED code is by definition also a t-BUED code. However, for a given value of r, burst unidirectional error-detecting codes can usually be found which can detect errors in a larger number of bits. Two systematic t-BUED codes have been presented in [BOSE86, BLAU88].

Bose code. Let $X = (x_1, x_2, .., x_k)$ be the given information symbol which has to be encoded. The check symbol CS is given by $CS \equiv k_0 \bmod 2^r$, where k_0 is the zero-count as before. Of course, it is assumed that $k \geq 2^r$, otherwise Berger code could be used to detect a unidirectional error in any number of bits. Let CS be denoted as $(y_1, y_2, .., y_r)$. The check symbol is not directly concatenated with the information symbol. Instead, the bits of the codeword are arranged in a particular format as follows:

$$x_1 x_2 .. x_{k-2^{r-1}} y_1 x_{k-2^{r-1}+1} .. x_{k-1} x_k y_2 y_3 .. y_r.$$

This code is 2^{r-1}-BUED.

Let $k = 13$, $r = 3$ and the information symbol $X = 1000010000100$. Since $k_0 = 10$, $CS = 10 \bmod 2^3 = 2$ (010 in binary). Therefore the bits will be arranged as shown in Fig. 6.3.

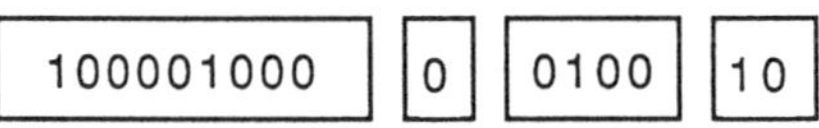

Fig. 6.3 Code format

No burst unidirectional error in four or fewer bits can change one codeword into another.

Other codes. Blaum has given a more efficient t-BUED code in [BLAU88], which, however, requires a slightly more complex encoding and decoding. For this code the value of t is given by

$$t = 2^r - \sum_{\alpha=0}^{r-1} \binom{\alpha}{\lfloor \alpha/2 \rfloor} .$$

A method for deriving a separable code based on the systematic t-UED and t-BUED codes presented in [BOSE85, JHA87a, BOSE86, BLAU88] has been given in [JHA89b]. This separable code detects a unidirectional error in at least as many, and frequently many more, bits as detected by the corresponding systematic code.

6.3 SELF-CHECKING CHECKERS

Many methods have been presented for designing gate-level checkers for the codes described earlier. Not as much, however, has been done in investigating the MOS implementations of self-checking circuits. Some early work in the area of nMOS implementations of self-checking circuits was reported in [JHA84, NICO85, JHA87b]. However, these techniques are not always directly applicable to CMOS implementations. In this section we will look at gate-level and CMOS designs for the checkers of some of the codes mentioned earlier. Both static and dynamic CMOS implementations of self-checking checker designs are possible. These are compared next.

6.3.1 Static vs Dynamic CMOS Implementations

The dynamic CMOS designs enjoy many advantages over the static CMOS designs of a checker [MANT84, JHA88]. Let us consider them next.

We know that stuck-on faults in CMOS circuits may require current monitoring. However, in the self-checking environment, current monitoring is not practical because it requires that test vectors be fed relatively slowly. In the case of dynamic CMOS circuits a stuck-on fault in the nMOS logic network of any gate does not require current monitoring. The need for current monitoring in the clocked transistors and inverters can be avoided by using the concept of p-dominance and n-dominance, as we will see later. Recall that a CMOS gate is p-dominant (n-dominant) if its output is 1(0) when both its networks conduct due to a fault. When these concepts are used, the presence of a stuck-on fault or fault sequence which remains undetected in a dynamic CMOS implementation of a self-checking circuit does not destroy the self-checking property. Therefore stuck-on faults can be included in the fault model for these self-checking circuits. In fact, a comprehensive fault model would include stuck-at, stuck-open and stuck-on faults. On the other hand, with only logic monitoring allowed for a static CMOS implementation of a self-checking circuit, many single stuck-on faults will remain undetected. When such an undetected fault is present, faults that occur later may destroy the self-checking property. Thus stuck-on faults can not be included in the fault model for static CMOS self-checking circuits.

The second advantage that dynamic CMOS implementations enjoy is that very few stuck-open faults in them require two-pattern tests. Also, these faults can be robustly detected. Thus no redesign for testability is needed. However, every stuck-open fault in a static CMOS implementation requires a two-pattern test. In the case of self-checking circuits the two vectors in the two-pattern tests must both be codewords. Frequently, a static CMOS implementation of a gate-level self-checking circuit is not robustly testable with two-pattern tests made up of only codewords. Thus redesign for testability is the rule rather than an exception.

Lastly, we know that certain assumptions have to be made for proper operation of self-checking circuits (see Section 6.1). One of these assumptions is that there is a sufficient time interval between any two faults so that all the required tests are applied to the circuit. In a large static CMOS self-checking circuit there will also be a large number of potential stuck-open faults, all of which require two-pattern tests made up of codewords. However, it may not always be realistic to assume that under normal operation the circuit receives all the two-pattern tests with the codewords ordered in the proper fashion. In the case of dynamic CMOS self-checking circuits the assumptions remain realistic because relatively much fewer two-pattern tests are required. However, in spite of the testability advantages enjoyed by dynamic CMOS circuits, static CMOS circuits continue to remain popular. Hence, static CMOS implementations of self-checking circuits can not be ignored.

6.3.2 Two-Rail Checkers

The n-variable two-rail checker has two rails of inputs $(x_1,x_2,..,x_n)$ and $(y_1,y_2,..,y_n)$, and two outputs (z_1,z_2). It is generally designed as a tree in which the nodes are any p-variable two-rail checker, $p < n$ [ANDE71]. The most common choice is to use $p = 2$.

We denote a p-variable two-rail checker by TRC_p. The gate-level design of TRC_2 is shown in Fig. 6.4. The 2-variable two-rail code has only four codewords: $(x_1,y_1,x_2,y_2) = 0101$, 0110, 1001 and 1010. For these codewords the outputs (z_1,z_2) of the checker have (0,1) or (1,0) which can be thought of as a 1-variable two-rail code or, equivalently, a 1-out-of-2 code. All the other vectors result in (0,0) or (1,1) at the outputs. Thus TRC_2 is code-disjoint. The four codewords also detect all single stuck-at faults in the checker. Thus it is self-testing. Also, no single fault can change a (0,1) at the output to (1,0) or *vice versa*, when a codeword is fed to the checker. Therefore the checker is fault-secure.

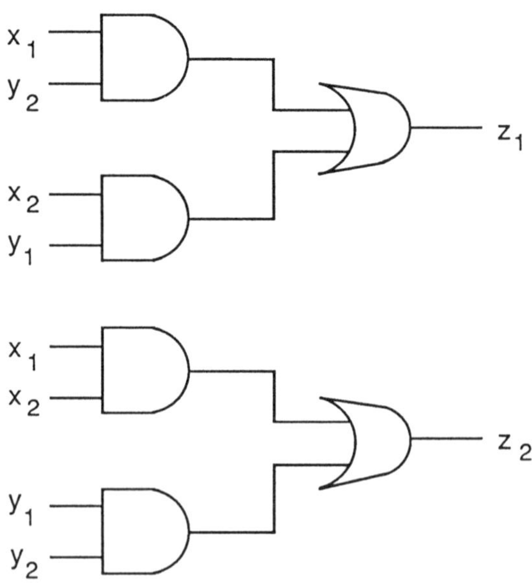

Fig. 6.4 Gate-level design for TRC_2

We can thus conclude that the checker is totally self-checking. Since there are no inverters in the circuit we can also conclude that the checker is totally self-checking with respect to all unidirectional stuck-at faults. Recall that a unidirectional stuck-at fault is a multiple stuck-at fault all whose constituent faults are of the same type.

A design for TRC_4 using TRC_2 as building blocks is shown in Fig. 6.5. It was shown in [ANDE71] that in the design of TRC_n, if the biggest building block in the tree is TRC_p, then a test set consisting of 2^p codewords can be found which detects all single and unidirectional stuck-at faults in TRC_n. Thus if only TRC_2 is used as a building block then only four codeword tests are needed for

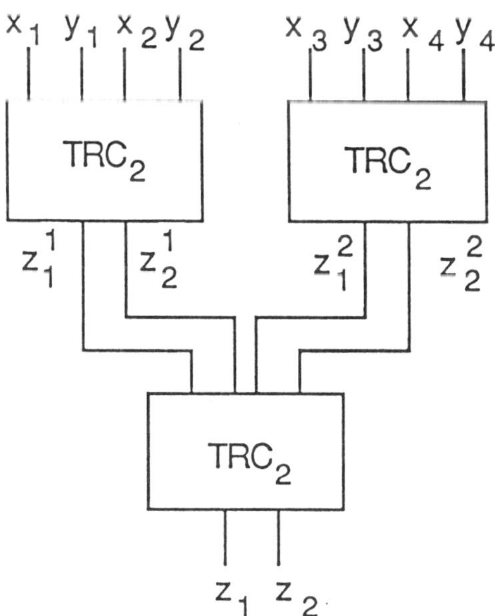

Fig. 6.5 A design for TRC_4

TRC_n irrespective of the size of the tree.

Dynamic CMOS implementations. The gate-level two-rail checker can be implemented using both the domino CMOS and DCVS techniques [JHA88, JHA89a]. The domino CMOS implementation of the gate-level TRC_2 is given in Fig. 6.6 and the DCVS implementation in Fig. 6.7. We can see that the two nMOS logic networks in the domino CMOS implementation have been merged into one in the DCVS implementation. Since we want to avoid current monitoring in a self-checking environment, we need some design constraints for both the domino CMOS and DCVS implementations. These constraints involve making the complex gate p-dominant and the inverter either n-dominant or p-dominant. To make the complex gate p-dominant, we may have to increase the width of the clocked pMOS transistors. This reduces the resistance of the transistor. Therefore when both nMOS and pMOS networks of the complex gate conduct due to a

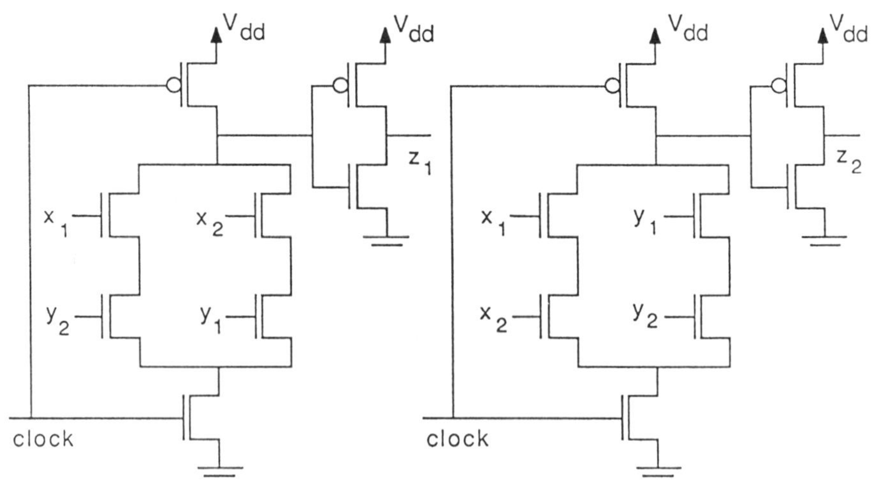

Fig. 6.6 Domino CMOS implementation of TRC_2

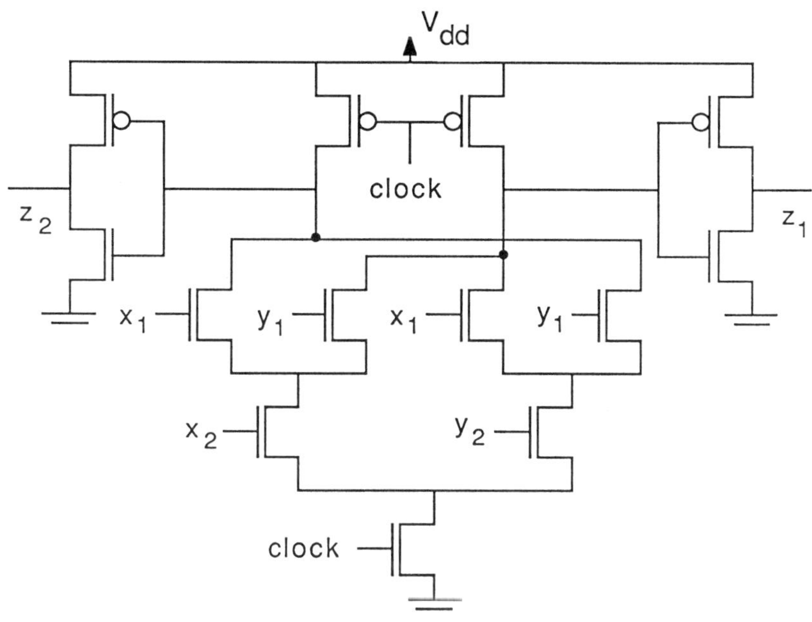

Fig. 6.7 DCVS implementation of TRC_2

fault, we can ensure that the output node of the complex gate has logic 1. Similarly, an inverter can be made n-dominant (p-dominant) by increasing the width of its nMOS (pMOS) transistor, if necessary.

Suppose that the complex gates and inverters in the two implementations have been made p-dominant and n-dominant respectively. Consider the set of codewords $(x_1,y_1,x_2,y_2) = \{0101,$ 0110, 1001, 1010\}. These codewords, if applied in the given sequence, will detect all detectable single stuck-open, stuck-on and stuck-at faults in either implementation (see Chapter 3). However, the stuck-on faults in the clocked nMOS transistors and the pMOS transistors of the inverters remain undetectable. But these faults or a sequence of these faults do not change the logical behavior of the circuit. They also do not prevent the detection of any other kind of single fault that may occur later. Lastly, no single fault in the circuit can change a (1,0) at (z_1,z_2) to (0,1) or *vice versa*. Therefore the two implementations are either (a) self-testing and fault-secure, or (b) fault-secure and continue to map codewords (non-codewords) at the inputs to codewords (non-codewords) at the outputs. When the next fault occurs, either property (a) or (b) is true for the fault sequence. In other words, these implementations are strongly self-checking.

A dynamic CMOS n-variable two-rail checker can be designed as before as a tree in which the nodes are any dynamic CMOS p-variable two-rail checker, $p < n$. For a DCVS implementation it was shown in [JHA89a] that at most only five codewords can detect all the detectable single faults in the tree, irrespective of the number of primary inputs or the number of inputs to the largest two-rail checker node in the tree. In some cases only four codeword tests are required. It is very advantageous to design easily testable checkers which require a small number of codeword tests. Firstly, it is easier to meet the assumptions made for self-checking circuits (see Section 6.1). Secondly, after an error is indicated by the checker, it becomes necessary to apply a diagnostic test to verify if the error was caused by a permanent fault, not a transient fault [ANDE71, MARO78a]. The time taken for this verification is reduced if the checker is easily testable, thereby improving the system performance.

6.3.3 Parity Checkers

Self-checking parity checkers can be designed for both odd-parity and even-parity codes [CART68]. Let $(x_1,x_2,..,x_{n+1})$ denote the inputs to the checker, where x_{n+1} corresponds to the parity bit. These inputs are divided into two disjoint groups with each group having at least one input. Two parity circuits are then realized, one for each group. A parity circuit can be implemented as a tree of EX-OR gates. For an odd-parity code the two parity circuits constitute a checker. However, in the checker for an even-parity code, the output of one of the parity circuits is inverted.

An odd-parity checker with seven inputs is shown in Fig. 6.8. A single fault in the checker can only affect one of the outputs. Thus the checker is fault-secure. It was shown in [ANDE71] that a parity checker in which the largest EX-OR gate has p inputs can be tested by 2^p codewords. Testing of parity trees has also been discussed in [BOSS70]. For the parity checker in Fig. 6.8 only four codeword tests are required. One possible set of tests is {0000001, 1110110, 1011101, 0101010}. Therefore the checker is self-testing. Finally, the checker can also be easily seen to be code-disjoint.

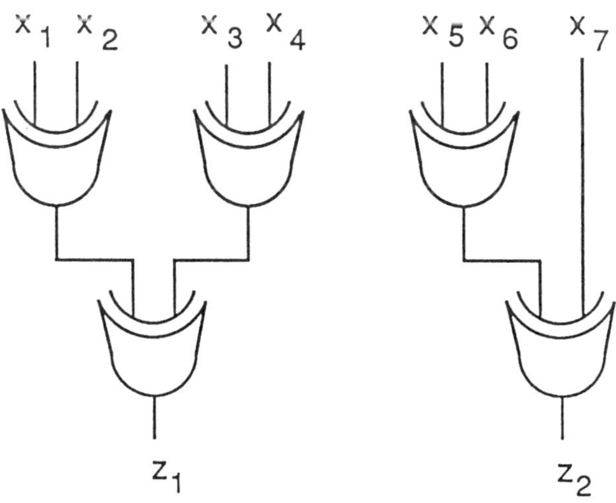

Fig. 6.8 An odd-parity checker

Hence, it is totally self-checking.

Cascode voltage switch implementation. A cascode voltage switch (CVS) implementation can be directly obtained from the gate-level parity checker [JHA89a]. For example, for the checker in Fig. 6.8, the CVS implementation is as shown in Fig. 6.9. DCVS and SCVS EX-OR gates have been discussed earlier in Sections 1.3.2 and 3.6. We again need to use the design constraints that were mentioned in Section 6.3.2. This involves making each complex gate in the checker p-dominant and making each inverter either n-dominant or p-dominant. It was shown in [JHA89a] that the CVS implementation of a totally self-checking parity checker is strongly self-checking and that it requires at most only five codeword tests to detect all single detectable stuck-open, stuck-on and stuck-at faults, irrespective of the number of primary inputs or the number of inputs to the largest EX-OR gate in the

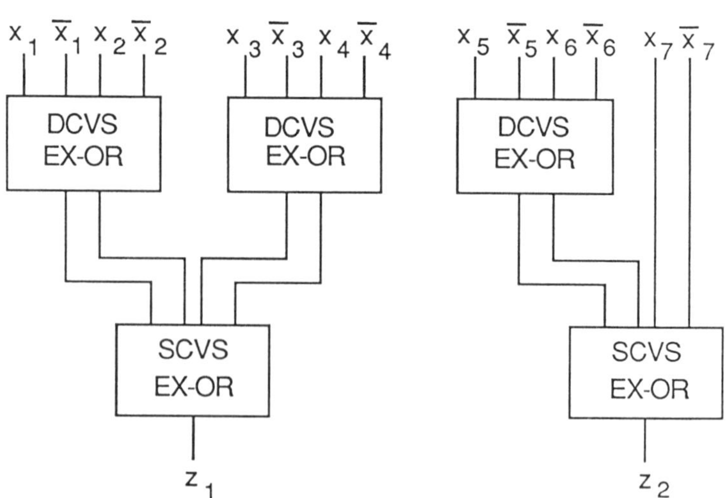

Fig. 6.9 The CVS implementation of a parity checker

checker.

6.3.4 m-out-of-n Checkers

Many gate-level totally self-checking checker designs have been presented for m-out-of-n codes [ANDE73, REDD74, SMIT77a, MARO78b, IZAW81, EFST83, NANY83, PIES83, GAIT83]. An m-out-of-n code in which m is equal or nearly equal to half of n is very popular since it contains the most number of codewords for the given n.

The first design for the m-out-of-2m totally self-checking checker was presented in [ANDE73]. Let the 2m bits be divided into two groups $A = \{x_1, x_2, .., x_m\}$ and $B = \{x_{m+1}, x_{m+2}, .., x_{2m}\}$, and let m_a and m_b denote the number of 1s in groups A and B respectively. The two output functions of the checker are given as

$$z_1 = \sum_{i=0}^{m} T(m_a \geq i) \cdot T(m_b \geq m{-}i) \quad \text{i odd}$$

$$z_2 = \sum_{i=0}^{m} T(m_a \geq i) \cdot T(m_b \geq m{-}i) \quad \text{i even}$$

where $T(m_a \geq i)$ is a threshold function which is 1 if and only if m_a is greater than or equal to i. The function $T(m_b \geq m{-}i)$ has similar meaning.

Suppose we want to design a checker for the 2-out-of-4 code. Then

$$z_1 = T(m_a \geq 1) \cdot T(m_b \geq 1)$$

$$= (x_1 + x_2) \cdot (x_3 + x_4)$$

$$z_2 = T(m_a \geq 0) \cdot T(m_b \geq 2) + T(m_a \geq 2) \cdot T(m_b \geq 0)$$

$$= 1 \cdot x_3x_4 + x_1x_2 \cdot 1$$

$$= x_1x_2 + x_3x_4$$

The gate-level design is shown in Fig. 6.10. Since no inverters are needed in the design, the checker is totally self-checking with respect to not only single but unidirectional stuck-at faults as well. The number of codeword tests required for the checker is 2^m.

A method was also suggested in [ANDE73] for designing a totally self-checking checker for m-out-of-n codes for which $n \neq 2m$. In this method the given code is first translated into a 1-out-of-$\binom{n}{m}$ code, which in turn is translated into a k-out-of-2k

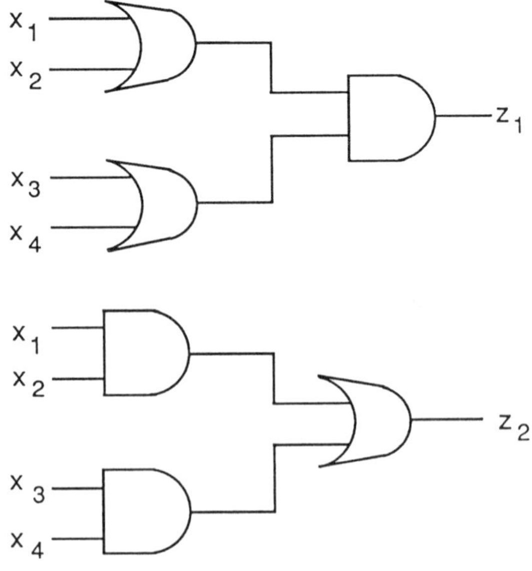

Fig. 6.10 A 2-out-of-4 checker

code. A checker for the k-out-of-2k code can be designed by the method presented earlier. Many other totally self-checking checker designs also exist for m-out-of-n codes (see references at the beginning of the section). The only code for which no gate-level totally self-checking checker has been found is the 1-out-of-3 code.

Domino CMOS implementations. All known gate-level totally self-checking checkers for m-out-of-n codes are inverter-free. The checkers are made up of AND and OR gates only. Therefore a domino CMOS implementation of these checkers is always possible. If the design constraints mentioned in previous sections are satisfied then the domino CMOS implementation can be shown to be strongly self-checking [JHA88].

A domino CMOS implementation of the 2-out-of-4 checker in Fig. 6.10 is given in Fig. 6.11. This implementation is strongly self-checking with respect to stuck-open, stuck-on and stuck-at

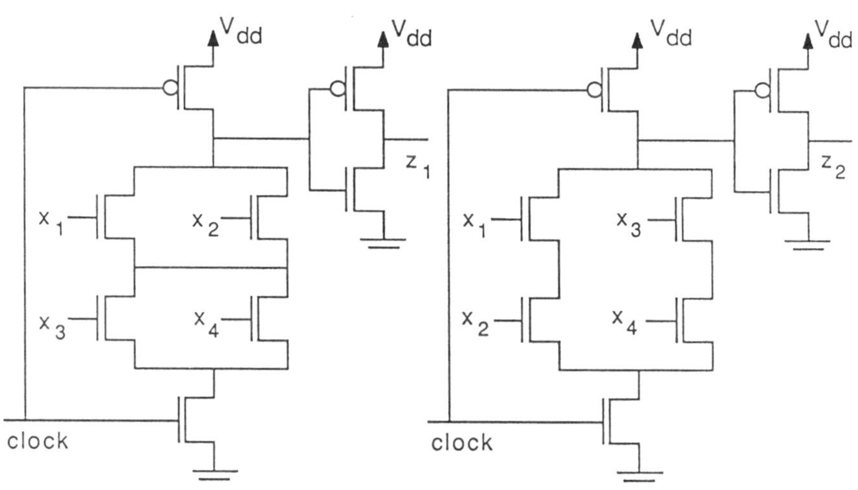

Fig. 6.11 Domino CMOS implementation of the
2-out-of-4 checker

faults.

A static CMOS m-out-of-n totally self-checking checker design is given in [KUND89]. Such a checker, however, can not be obtained directly from the gate-level design. Also, as explained in Section 6.3.1, the stuck-on faults can not be included in the fault model for a static CMOS checker.

It is worth pointing out that an nMOS checker has been found for the 1-out-of-3 code which is totally self-checking with respect to realistic physical failures like gate-drain short, gate-source short, gate-substrate short, drain-contact open, source-contact open and floating gate [TAO88, JHA89c]. However, a CMOS totally self-checking checker for the 1-out-of-3 code has not been found.

6.3.5 Berger Checkers

The Berger code is an optimal all-unidirectional error-detecting systematic code. The most common totally self-checking checker design for systematic codes is shown in Fig. 6.12 [ASHJ77, MARO78a].

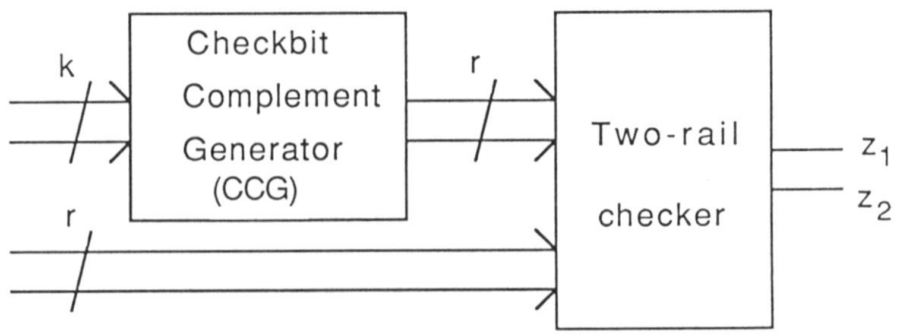

Fig. **6.12** Totally self-checking checker for systematic codes

The k information bits are fed to a checkbit complement generator (CCG) which generates the bitwise complement of the checkbits.

A two-rail checker verifies if the output bits of the CCG are in fact complement of the r received checkbits. If CCG is irredundant and the two-rail checker receives all its tests then the checker is totally self-checking. However, such a checker may require all the 2^k codeword tests. In order to make the checker easily testable we should not use any arbitrary design for the CCG. For Berger codes an easily testable gate-level checker has been given in [MARO78a]. In this design the CCG is made up of an interconnection of full and half-adders. For a checker for which $k \neq 2^{r-1}$, this design requires at most $8r - 3$ codeword tests. When $k = 2^r - 1$, the design requires at most 12 codeword tests. No totally self-checking checker design for a Berger checker with two outputs has been discovered for the case $k = 2^{r-1}$. However, in this case a totally self-checking checker can be obtained for an equivalent code [ASHJ77].

Although no domino CMOS implementation of the Berger checker is possible, a CVS implementation is. However, such an implementation has not been evaluated before.

6.3.6 Checkers for Borden, Bose-Lin, Bose and Blaum Codes

These codes are t-unidirectional error-detecting (t-UED) or t-burst unidirectional error-detecting (t-BUED). The Borden code is less redundant than the m-out-of-n code whereas the Bose-Lin, Bose and Blaum codes are less redundant than the Berger code. Thus self-checking functional circuits and checkers based on these codes can be expected to occupy less area on the chip. This fact ought to result in increased popularity for these and other t-UED codes in the future.

A totally self-checking checker design for the Borden code has been presented in [JHA89d]. Totally self-checking checker designs for Bose-Lin, Bose and Blaum codes have been presented in [JHA89e]. The checkers for Bose-Lin and Bose codes require only $O(\log_2 k)$ codeword tests whereas the checker for the Blaum code requires $O(k)$ codeword tests, where k is the number of information bits. Thus these checker designs are easily testable. A checker design for the Bose code also appears in [BOSE86]. However, this design has not been shown to be easily testable. CMOS implementations of any of these designs have not yet been evaluated.

6.3.7 Embedded Checker Problem

A checker is said to be embedded if the set of codewords it receives from the functional circuit under normal operation is not enough to detect all detectable faults from the fault model. This can prevent the checker from being self-checking. This situation can arise when the functional circuit generates only a subset of its output code space. This problem is not severe for easily testable checkers because such checkers require a very small subset of the code space for self-testing purposes. Thus the available codewords should usually be successful in flushing out all the detectable faults. This is one more advantage of easily testable checkers, which was not pointed out earlier. However, in general, one may encounter embedded checkers.

This problem was first attacked in [KHAK82, KHAK84]. In these papers it was shown how embedded parity trees and checkers can be made self-testing with available vectors by interchanging input lines. These methods work if some simple conditions are satisfied. In [JHA89f] a method has been given for designing sufficiently strongly self-checking (SSSC) embedded checkers for any systematic or separable code. Such a checker continues to operate properly even if it has a fault which has not been exposed due to unavailability of codewords from the functional circuit. In fact the checker performs correctly for most sequences of faults as well. Another approach to this problem is to periodically test the checker by feeding it vectors from a test pattern generator [NICO84, NICO89].

6.4 SELF-CHECKING FUNCTIONAL CIRCUITS

The design of totally self-checking functional circuits was presented in [SMIT77b] and in more detail in [SMIT76]. In [SMIT78] it was shown how strongly fault-secure functional circuits can be designed. Totally self-checking and strongly fault-secure programmable logic array (PLA) designs for functional circuits were discussed in [WANG79] and [MAK82] respectively. A comprehensive theory for the design of a totally self-checking system was presented in [SMIT83].

The totally self-checking circuit designs presented in [SMIT76] are shown to be inverter-free if they are based on unordered (all-unidirectional error-detecting) codes. Therefore a domino CMOS implementation of these circuits is possible. Given the design constraints mentioned earlier, the domino CMOS implementations can be shown to be strongly fault-secure with respect to the stuck-open, stuck-on and stuck-at faults [JHA88].

REFERENCES

[ANDE71] D. A. Anderson, "Design of self-checking digital networks using coding techniques," Tech. Report R-527, CSL, Univ. of Illinois, IL, 1971.

[ANDE73] D. A. Anderson and G. Metze, "Design of totally self-checking circuits for m-out-of-n codes," *IEEE Trans. Comput.*, vol. C-22, pp. 263-269, Mar. 1973.

[ASHJ77] M. J. Ashjaee and S. M. Reddy, "On totally self-checking checkers for separable codes," *IEEE Trans. Comput.*, vol. C-26, pp. 737-744, Aug. 1977.

[AVIZ71] A. Avizienis, "Arithmetic codes: Cost and effectiveness studies for application in digital system design," *IEEE Trans. Comput.*, vol. C-20, pp. 1322-1331, Nov. 1971.

[BERG61] J. M. Berger, "A note on error detection codes for asymmetric channels," *Inform. Contr.*, vol. 4, pp. 68-73, Mar. 1961.

[BLAU88] M. Blaum, "On systematic burst unidirectional error detecting codes," *IEEE Trans. Comput.*, vol. 37, pp. 453-457, Apr. 1988.

[BORD82] J. M. Borden, "Optimal asymmetric error detecting codes," *Inform. Contr.*, vol. 53, pp. 66-73, Apr. 1982.

[BOSE82] B. Bose and T. R. N. Rao, "Theory of unidirectional error correcting/detecting codes," *IEEE Trans. Comput.*, vol. C-31, pp. 521-530, June 1982.

[BOSE85] B. Bose and D. J. Lin, "Systematic unidirectional error-detecting codes," *IEEE Trans. Comput.*, vol. C-34, pp. 1026-1032, Nov. 1985.

[BOSE86] B. Bose, "Burst unidirectional error detecting codes," *IEEE Trans. Comput.*, vol. C-35, pp. 350-353, Apr. 1986.

[BOSS70] D. C. Bossen, D. L. Ostapko, and A. M. Patel, "Optimum test patterns for parity networks," in *Proc. AFIPS Conf.*, Houston, TX, pp. 63-68, Nov. 1970.

[CART68] W. C. Carter and P. R. Schneider, "Design of dynamically checked computers," in *Proc. IFIP '68*, Edinburgh, Scotland, pp. 878-883, Aug. 1968.

[DONG82] H. Dong, "Modified Berger codes for detection of unidirectional errors," in *Proc. Int. Symp. Fault-Tolerant*

Comput., Santa Monica, pp. 317-320, June 1982.

[EFST83] C. Efstathiou and C. Halatsis, "Modular realization of totally self-checking checkers for m-out-of-n codes," in *Proc. Int. Symp. Fault-Tolerant Comput.*, Milan, Italy, pp. 154-161, June 1983.

[FRIE62] C. V. Frieman, "Optimal error detection codes for completely asymmetric binary channels," *Inform. Contr.*, vol. 5, pp. 64-71, Mar. 1962.

[GAIT83] N. Gaitanis and C. Halatsis, "A new design method for m-out-of-n TSC checkers," *IEEE Trans. Comput.*, vol. C-32, pp. 273-283, Mar. 1983.

[IZAW81] N. Izawa, "3-level realization of self-checking 1-out-of-n code checkers," *1981 IECE Nat. Conv. Inf. Syst., Inst. Electron. Commun. Eng. Jap.*, no. 504, Oct. 1981 (in Japanese).

[JHA84] N. K. Jha and J. A. Abraham, "Totally self-checking MOS circuits under realistic physical failures," in *Proc. Int. Conf. Computer Design*, Port Chester, NY, pp. 665-670, Oct. 1984.

[JHA87a] N. K. Jha and M. B. Vora, "A systematic code for detecting t-unidirectional errors," in *Proc. Int. Symp. Fault-Tolerant Comput.*, Pittsburgh, PA, pp. 96-101, July 1987.

[JHA87b] N. K. Jha and J. A. Abraham, "Techniques for efficiently implementing totally self-checking checkers in MOS technology," *Int. J. Comp.& Math. with Applications*, vol. 13, pp. 555-566, 1987.

[JHA88] N. K. Jha, "SFS/SSC domino-CMOS implementations of TSC circuits," in *Proc. Annual Allerton Conf. Communication, Control & Computing*, Allerton, IL, pp. 768-777, Sept. 1988.

[JHA89a] N. K. Jha, "Fault detection in CVS parity trees: Application to SSC CVS parity and two-rail checkers," in *Proc. Int. Symp. Fault Tolerant Comput.*, Chicago, IL, pp. 107-114, June

1989.

[JHA89b] N. K. Jha, "Separable codes for detecting unidirectional errors," *IEEE Trans. CAD*, vol. 8, pp. 571-574, May 1989.

[JHA89c] N. K. Jha, "Comments on 'A MOS implementation of totally self-checking checker for the 1-out-of-3 code'," *IEEE J. Solid-State Circuits*, vol. 24, Oct. 1989.

[JHA89d] N. K. Jha, "A totally self-checking checker for Borden's code," *IEEE Trans. CAD*, vol. 8, pp. 731-736, July 1989.

[JHA89e] N. K. Jha, "Design of totally self-checking checkers for Bose-Lin, Bose and Blaum codes," in *Proc. Midwest Symp. Circuits & Systems*, Urbana, IL, Aug. 1989.

[JHA89f] N. K. Jha, "Design of sufficiently strongly self-checking embedded checkers for systematic and separable codes," in *Proc. Int. Conf. Computer Design*, Boston, Mass., Oct. 1989.

[KHAK82] J. Khakbaz, "Self-testing embedded parity trees," in *Proc. Int. Symp. Fault-Tolerant Comput.*, Santa Monica, pp. 109-116, June 1982.

[KHAK84] J. Khakbaz and E. J. McCluskey, "Self-testing embedded parity checkers," *IEEE Trans. Comput.*, vol. C-33, pp. 753-756, Aug. 1984.

[KUND89] S. Kundu and S. M. Reddy, "Design of TSC checkers for implementation in CMOS technology," in *Proc. Int. Conf. Computer Design*, Boston, Mass., Oct. 1989.

[MAK82] G. P. Mak, J. A. Abraham and E. S. Davidson, "The design of PLAs with concurrent error detection," in *Proc. Int. Symp. Fault-Tolerant Comput.*, Santa Monica, pp. 303-310, June 1982.

[MANT84] S. R. Manthani and S. M. Reddy, "On CMOS totally self-checking circuits," *Int. Test Conf.*, Philadelphia, PA, pp. 866-877, Oct. 1984.

[MARO78a] M. A. Marouf and A. D. Friedman, "Design of self-checking checkers for Berger codes," in *Proc. Int. Symp. Fault-

Tolerant Comput., Toulouse, France, pp. 179-184, June 1978.

[MARO78b] M. A. Marouf and A. D. Friedman, "Efficient design of self-checking checker for any m-out-of-n code," *IEEE Trans. Comput.*, vol. C-27, pp. 482-490, June 1978.

[NANY83] T. Nanya and Y. Tohma, "A 3-level realization of totally self-checking checkers for m-out-of-n codes," in *Proc. Int. Symp. Fault-Tolerant Comput.*, Milan, Italy, pp. 173-176, June 1983.

[NICO84] M. Nicolaidis, I. Jansch and B. Courtois, "Strongly code-disjoint checkers," in *Proc. Int. Symp. Fault-Tolerant Comput.*, Orlando, FL, pp. 16-21, June 1984.

[NICO85] M. Nicolaidis and B. Courtois, "Layout rules for the design of self-checking circuits," in *Proc. VLSI Conf.*, Tokyo, Aug. 1985.

[NICO89] M. Nicolaidis, "A unified built-in self-test scheme: UBIST," *IEEE Trans. CAD*, vol. 8, pp. 203-218, Mar. 1989.

[PIES83] S. Piestrak, "Design method of totally self-checking checkers for m-out-of-n codes," in *Proc. Int. Symp. Fault-Tolerant Comput.*, Milan, Italy, pp. 162-168, June 1983.

[PRAD80a] D. K. Pradhan and J. J. Stiffler, "Error-correcting codes and self-checking circuits in fault-tolerant computers," *IEEE Computer*, pp. 27-37, Mar. 1980.

[PRAD80b] D. K. Pradhan, "A new class of error correcting/ detecting codes for fault-tolerant computer applications," *IEEE Trans. Comput.*, vol. C-29, pp. 471-481, June 1980.

[REDD74] S. M. Reddy, "A note on self-checking checkers," *IEEE Trans. Comput.*, vol. C-23, pp. 1100-1102, Oct. 1974.

[SMIT76] J. E. Smith, "The design of totally self-checking combinational circuits," Tech. Report R-737, CSL, Univ. of Illinois, IL, Aug. 1976.

[SMIT77a] J. E. Smith, "The design of totally self-checking check circuits for a class of unordered codes," *J. Des. Automat. &*

Fault-Tolerant Comput., pp. 321-343, Oct. 1977.

[SMIT77b] J. E. Smith and G. Metze, "The design of totally self-checking combinational circuits," in *Proc. Int. Symp. Fault-Tolerant Comput.*, Los Angeles, CA, pp. 130-134, June 1977.

[SMIT78] J. E. Smith and G. Metze, "Strongly fault-secure logic networks," *IEEE Trans. Comput.*, vol. C-27, pp. 491-499, June 1978.

[SMIT83] J. E. Smith and P. Lam, "A theory of totally self-checking system design," *IEEE Trans. Comput.*, vol. C-32, pp. 831-843, Sept. 1983.

[SMIT84] J. E. Smith, "On separable unordered codes," *IEEE Trans. Comput.*, vol. C-33, pp. 741-743, Aug. 1984.

[TAMI84] Y. Tamir and C. H. Sequin, "Design and application of self-testing comparators implemented with MOS PLAs," *IEEE Trans. Comput.*, vol. C-33, pp. 493-506, June 1984.

[TAO88] D. L. Tao, P. K. Lala, and C. R. P. Hartman, "A MOS implementation of totally self-checking checker for the 1-out-of-3 code," *J. Solid-State Circuits*, vol. 23, pp. 875-877, June 1988.

[WANG79] S. L. Wang and A. Avizienis, "The design of totally self-checking circuits using Programmable Logic Arrays," in *Proc. Int. Symp. Fault-Tolerant Comput.*, Madison, WI, pp. 173-180, June 1979.

ADDITIONAL READING

[BRUC89] J. Bruck and M. Blaum, "Some new EC/AUED codes," in *Proc. Int. Symp. Fault-Tolerant Comput.*, Chicago, IL, pp. 208-215, June 1989.

[BOSE82] B. Bose and D. K. Pradhan, "Optimal unidirectional error detecting/correcting codes," *IEEE Trans. Comput.*, vol. C-31, pp. 564-568, June 1982.

[LIN88] D. J. Lin and B. Bose, "Theory and design of t-error correcting and $d(d > t)$-unidirectional error detecting (t-EC d-UED) codes," *IEEE Trans. Comput.*, vol. 37, pp. 433-439, Apr. 1988.

[NIKO86] D. Nikolos, N. Gaitanis and G. Philokyprou, "Systematic t-error correcting/all unidirectional error detecting codes," *IEEE Trans. Comput.*, vol. C-35, pp. 394-402, May 1986.

[NIKO88] D. Nikolos, A. M. Paschalis and G. Philokyprou, "Efficient design of totally self-checking checkers for all low-cost arithmetic codes," *IEEE Trans. Comput.*, vol. C-37, pp. 807-814, July 1988.

[PASC88] A. M. Paschalis, D. Nikolos and C. Halatsis, "Efficient modular design of TSC checkers for m-out-of-2m codes," *IEEE Trans. Comput.*, vol. 37, pp. 301-309, Mar. 1988.

[TAO88] D. L. Tao, C. R. P. Hartman and P. K. Lala, "An efficient class of unidirectional error detecting/correcting codes," *IEEE Trans. Comput.*, vol. 37, pp. 879-882, July 1988.

PROBLEMS

6.1. The Berger code is an optimal systematic all-unidirectional error-detecting (AUED) code when all the 2^k possible k-bit information symbols have to be encoded. But it may not be optimal when only a proper subset of the set of all information symbols has to be encoded. Ascertain this fact by encoding the following set I of information symbols into an AUED code by using fewer checkbits than the Berger code would require.

$$I = \{00011, 00110, 01110, 10001, 10110, 11100\}$$

6.2. In Prob. 6.1 if the information symbol 00110 in I is replaced by 01001 then what is the least number of checkbits you would require to encode the information symbols into an AUED code? Show your encoding.

6.3. Bose and Lin presented two methods for deriving t-unidirectional error-detecting (t-UED) codes when the number of checkbits $r \geq 5$, as explained in Section 6.2.2. Show that for the code derived by Method 1, $t = 2^{r-2} + r - 2$, and for the code derived by Method 2, $t = 5 \times 2^{r-4} + r - 4$.

6.4. Prove that if a gate-level checker, which is made up of an interconnection of AND and OR gates, is self-testing with respect to all single stuck-at faults then it is also self-testing with respect to all unidirectional stuck-at faults.

6.5. A parity tree all whose nodes are 2-input EX-OR gates is called a binary parity tree. The following sequences, which are called cyclic diagnostic test sequences, can be used to test such trees: $A_0 = 0110$, $A_1 = 0011$, $A_2 = 0101$. One can check that $A_0 = A_1 \oplus A_2$, $A_1 = A_0 \oplus A_2$ and $A_2 = A_0 \oplus A_1$. Furthermore, any two test sequences together contain all possible 2-bit patterns: 00, 01, 10, 11. Using these test sequences show that any gate-level binary parity tree can be tested for all single stuck-at faults by only four vectors.

6.6. Give a design for a 6-variable gate-level two-rail checker using 2-variable two-rail checkers as building blocks. Show that all the single stuck-at faults in the design can be detected by a set of only four codewords. (Hint: use cyclic diagnostic test sequences from Prob. 6.5).

6.7. Extend the method developed in Prob. 6.5 to detect all single stuck-at faults in any binary odd-parity checker. You are allowed to use only codeword tests. (Hint: four codewords are enough).

6.8. Suppose that the outputs of a functional circuit have been encoded using an odd-parity code. The encoded word is of the form $\{x_1, x_2, x_3, x_4\}$. This code normally has eight codewords. But suppose that only the following codewords are available from the functional circuit: $\{0001, 0010, 1101, 1110\}$. Design a gate-level parity checker which is self-testing even if only these codewords are fed to it.

6.9 Suppose that a gate-level 4-variable two-rail checker only receives the following codewords: $(x_1, y_1, x_2, y_2, x_3, y_3, x_4, y_4) = \{01011001, 10100110, 10101001, 01010110\}$. Design the checker to be self-testing with only these codewords.

6.10. Consider the design of the m-out-of-n gate-level totally self-checking checker presented in Section 6.3.4. Show that

(a) If the checker receives a non-codeword with $(m+1)$ or more 1s then $(z_1, z_2) = (1,1)$.

(b) If the checker receives a non-codeword with $(m-1)$ or fewer 1s then $(z_1, z_2) = (0,0)$.

6.11. Design a gate-level 3-out-of-6 totally self-checking checker. Derive a minimal set consisting of only codewords, which detects all single stuck-at faults in the design.

6.12. Consider the domino CMOS implementation of the design obtained in Prob. 6.11. Derive a minimal test set of codeword tests which detects all detectable single stuck-open, stuck-on and stuck-at faults in it. Show that this implementation is strongly self-checking.

6.13. Derive a minimal test set of codewords to detect all detectable single stuck-open, stuck-on and stuck-at faults in the CVS implementation given in Fig. 6.9. Show that this implementation is strongly self-checking.

Chapter 7
CONCLUSIONS

In the previous chapters we have considered testing, design for testability and self-checking design of combinational static and dynamic CMOS circuits. Not much work has been done in the area of test generation for sequential CMOS circuits. Owing to the difficult and time-consuming nature of sequential circuit testing, usually only the stuck-at fault model is assumed even for CMOS circuits. In order to reduce the complexity of sequential circuit testing, design for testability techniques, such as scan techniques, are used. These techniques reduce the complexity of test generation to that of testing combinational circuits. For example, in the scan path technique [WILL73], the circuit has a normal mode of operation and a test mode. In the test mode the circuit flip-flops are interconnected to form a shift register (i.e. the scan path). It is easy to test such a scan path. The test patterns for the combinational part of the circuit are serially shifted in and the responses are serially shifted out through the flip-flops. A scan path technique called level-sensitive scan design (LSSD) was developed for latch-based systems [EICH77], and is extensively used in IBM. Some chip area penalty has to be paid to incorporate these design for testability features into the chip.

Even when scan techniques are used, test generation is only

geared towards stuck-at faults. The reason is that with the basic scheme it is not possible to first initialize the circuit and then test it for stuck-open faults, because in the process of shifting in the test vector the initialization can be destroyed. However, a technique was introduced in [LIU87] which allows a sequential CMOS circuit employing scan path to be robustly tested for stuck-open faults. This is accomplished by using a modified shift register latch design. A discussion on stuck-open fault detection in CMOS memory elements can be found in [REDD85].

Another area of testing which has attracted a lot of attention is built-in self-test (BIST). In this scheme the test patterns are generated on the chip itself by a test pattern generator. The response from the circuit is compressed into a signature. After all the test patterns are fed to the circuit, the signature which is obtained is compared to the reference signature to determine if a fault is present in the circuit. However, the basic scheme is not designed to detect stuck-open faults. An extension of this scheme to detect stuck-open faults in both sequential and combinational CMOS circuits was given in [STAR84]. A BIST approach called pseudo-exhaustive adjacency test (PEAT) was proposed in [CRAI85] for detecting stuck-open and stuck-at faults in CMOS circuits. However, more work needs to be done for adapting the testing techniques, such as scan path and BIST, to CMOS circuits.

REFERENCES

[CRAI85] G. L. Craig and C. R. Kime, "Pseudo-exhaustive adjacency test: A BIST approach for stuck-open faults," in *Proc. Int. Test Conf.*, Philadelphia, PA, pp. 126-137, Oct. 1985.

[EICH77] E. B. Eichelberger and T. W. Williams, "A logic design structure for LSI testability," in *Proc. Design Automation Conf.*, New Orleans, pp. 462-468, June 1977.

[LIU87]   D. L. Liu and E. J. McCluskey, "CMOS scan-path IC design for stuck-open fault testability," *IEEE J. Solid-State Circuits*, vol. SC-22, pp. 880-885, Oct. 1987.

[REDD85]   M. K. Reddy and S. M. Reddy, "On FET stuck-open fault detectable CMOS memory elements," in *Proc. Int. Symp. Fault-Tolerant Comput.*, Ann Arbor, MI, pp. 424-429, June 1985.

[STAR84]   C. W. Starke, "Built-in test for CMOS circuits," in *Proc. Int. Test Conf.*, Philadelphia, PA, pp. 309-314, Oct. 1984.

[WILL73]   M. J. Y. Williams and A. B. Angel, "Enhancing testability of large scale integrated circuits via test points and additional logic," *IEEE Trans. Comput.*, vol. C-22, pp. 46-60, Jan. 1973.